Lipid Mediators in Eye Inflammation

New Trends in Lipid Mediators Research

Vol. 5

Series Editor
Pierre Braquet, Le Plessis Robinson

KARGER

Basel · München · Paris · London · New York · New Delhi · Bangkok · Singapore · Tokyo · Sydney

Lipid Mediators in Eye Inflammation

Volume Editor
Nicolas G. Bazan, New Orleans, La.

38 figures and 18 tables, 1990

KARGER

Basel · München · Paris · London · NewYork · New Delhi · Bangkok · Singapore · Tokyo · Sydney

Nicolas G. Bazan, MD, PhD
LSU Eye Center and Neuroscience Center,
Louisiana State University Medical Center, School of Medicine,
New Orleans, La. (USA)

Library of Congress Cataloging-in-Publication Data
Lipid mediators in eye inflammation / volume editor, Nicolas G. Bazan
(New trends in lipid mediators research; vol. 5)
Includes bibliographical references.
1. Eye – Inflammation. 2. Inflammation – Mediators. 3. Lipids – Pathophysiology.
I. Bazan, Nicolas G. II. Series.
[DNLM: 1. Eye Diseases – drug therapy. 2. Lipids – metabolism.
W1NE513FN v. 5 / WW 166 L764]
RE96.L56 1990
617.7′1 – dc20
ISBN 3-8055-5067-7

Drug Dosage

The authors and the publisher have exerted every effort to ensure that drug selection and dosage set forth in this text are in accord with current recommendations and practice at the time of publication. However, in view of ongoing research, changes in government regulations, and the constant flow of information relating to drug therapy and drug reactions, the reader is urged to check the package insert for each drug for any change in indications and dosage and for added warnings and precautions. This is particularly important when the recommended agent is a new and/or infrequently employed drug.

Printed in Switzerland by Thür AG Offsetdruck, Pratteln
ISBN 3-8055-5067-7

Contents

Foreword

I believe that the work done by the authors in this book represents the kind of revolution in our intellectual understanding of inflammation that the storming of the Bastille represented in our search for freedom and democracy.

In the old days not so long ago the term inflammation was thought to have a specific defined meaning. It was considered as a generic term with little consideration of the wide variety of inciting factors and the wide variety of clinical and laboratory settings. Scientists and physicians were even so naive as to assume that the term 'anti-inflammatory' was defined and meaningful. The therapeutic implication in those days gone by was that 'anti-inflammatory agents' would all somehow act with the same non-specific mechanism of action. Since those 'dark ages' that were only two or three decades ago there has been a revolution in the understanding of the multiple facets of inflammation, the numerous mediators responsible for it, and the diversity of mechanisms by which it occurs. As little as we know about specific inflammatory mediators, the nonsteroidal cyclooxygenase inhibitors have revolutionized the treatment of arthritis, and a wide variety of ocular and nonocular diseases. Lipoxygenase inhibitors are bound to be similarly useful. Drugs such as sodium cromoglycolate that specifically inhibit mast cell degranulation have been useful in the treatment of asthma and a variety of other allergic conditions. Calcium channel blockers have been shown to reduce the tissue damage of anoxia and stroke, as have inhibitors of inflammatory mediators. The mediators of inflammation themselves sometimes produce useful and valuable selective tissue activity such as uterine contraction, and even lower intraocular pressure, whereas drugs such as ciclosporin by inhibiting helper cell activity in transplant rejection have created a new era in organ transplant safety that was never possible by less selective steroid use alone.

Our knowledge is still rudimentary and fragmentary. The remarkable thing to me is how much we have been able to accomplish in such a short time with such an incomplete understanding of the details of injury and inflammation. As we dissect out the mediators of specific responses, on the other hand, we are offered a way to treat undesirable activity in a far more effective and far safer way than ever before. Conversely, as these mediators and their activity become familiar to us, it will be possible to choose desirable functions and selectively turn them on and off without the kind of nonspecific tissue damage and destruction that might evolve were the whole cascade to be activated. In the eye, for example, there are instances in which we would like to induce vascularization, to raise or lower intraocular pressure selectively, to stimulate local immune defenses that are specific without some of the less selective components being involved. We would like to turn off undesirable consequences such as cystoid macular edema without turning off the tissue-healing responses to injury which are so necessary.

At this point in time my imagination like yours must be limited by the newness of this field and the amount of knowledge still to be acquired. If you and I read this book and decide that knowledge is incomplete, and the possible beneficial uses of this knowledge have just begun to be explored, take heart with me that this work clearly outlines the mechanisms by which such knowledge can be obtained, and the enormous potential benefits that will ultimately result from it.

Herbert E. Kaufman
MD, Boyd Professor of Ophthalmology and
Pharmacology and Experimental Therapeutics
Director, Louisiana State University Eye Center
Head, Department of Ophthalmology

Preface

The ocular inflammatory response involves an array of chemical mediators, among which lipid mediators comprise a structurally heterogeneous group of biologically active metabolites with diverse and potent effects. All lipid mediators share a common origin, either as part of, or synthesized from, a parent membrane phospholipid molecule. These preexisting phospholipids lack the biological actions of lipid mediators.

Lipid mediators of inflammation are also modulators (second messengers) of cellular function when generated under physiological conditions through receptor activation. One other common feature of lipid mediators is that either phospholipase A_2 or C catalyzes the initial step in the hydrolysis of the precursor, biologically inactive phospholipid.

The overall area of lipid mediators has expanded greatly in the last few years. A reflection of this growth has been the establishment of new journals focusing specifically on messengers, cell signal transduction and mediators, ranging in scope from the physiological significance to the pathophysiological role of these metabolites.

In eye research, an increasing interest in lipid mediators has developed. This stems from the need to identify the chemical and cellular events in various forms of eye inflammation. Lipid mediators are a group of chemicals that, along with several nonlipid mediators (e.g. cytokines), make up networks in response to inflammation. The identification of inhibitors of the synthesis of specific lipid mediators as well as of receptor antagonists will open up new approaches for the clinical management of various forms of eye inflammation heretofore elusive to therapy. All the knowledge generated will create a new understanding of the pathophysiology of eye inflammation, particularly because it will allow us to single out events and mediators that may couple the inflammatory response to the turning-on of wound healing at the genomic level. This in turn will lead to an effective treatment of the ocular wound-healing responses. This book represents a planned effort to gather themes focused on lipid mediators in eye inflammation. This book is not the Proceedings of a meeting. A synthesis of knowledge in this area is necessary because of rapidly expanding information and interest in this topic.

Many thanks to the contributing authors.

Nicolas G. Bazan

Bazan NG (ed): Lipid Mediators in Eye Inflammation.
New Trends Lipid Mediators Res. Basel, Karger, 1990, vol 5, pp 1–11

Response of Inflammatory Lipid Mediators following Corneal Injury[1]

Haydee E.P. Bazan

LSU Eye Center, Department of Ophthalmology, Louisiana State University School of Medicine, New Orleans, La., USA

During the past few years our research interests have focused on understanding how lipid mediators are involved in corneal inflammatory response. The deleterious consequences of inflammatory disease probably result from interactions between tissue-resident cells and infiltrate cells that arrive at the site of injury after the lesion forms.

Infiltrates of polymorphonuclear leukocytes (PMNs) have been found after simple trauma resulting from removal of the central corneal epithelium [1] and in corneal ulceration after alkali burn [2]. These cells can appear at the edge of the wound as early as 3 h after insult [1]. Corneal edema is induced by activated PMNs [3]. PMNs and mononuclear cells have been observed in the corneal stroma during herpetic corneal keratitis [4]. Accumulation of PMNs and macrophages at inflammation sites is required for the elimination of contaminating bacteria and debris in order to start the healing process. On the other hand, soon after corneal injury, depending on the degree of injury, the epithelium reorganizes and migrates over the wound [5], stroma keratocytes produce fibronectin [6], a glycoprotein with wound-healing properties [7], and endothelial cells take on a fibroblast-like morphology in order to migrate and cover the wound. Many biochemical changes occur during these processes. One of the first events is the formation of several mediators of inflammation as a consequence of the insult; among them, the metabolites of arachidonic acid (AA) are con-

[1] Portions of the work described herein, performed in the author's laboratory, were supported by United States Public Health Service Grants EY02377, EY04928, and EY06635, from the National Eye Institute, Bethesda, Md.

sidered very important lipid mediators of inflammation. The formation of cyclooxygenase metabolites of AA was first described in cultured corneal epithelial and fibroblast cells [8].

Models of Corneal Injury

In vitro Models

Several types of injury have been applied to the cornea of the rabbit eye in order to measure different lipid mediators. In the in vitro models, the injury is produced in the anesthetized cornea, and at times specified in each study the animals are killed and the eyes removed. Eight-millimeter central corneal trephine buttons are obtained, and either the entire tissue or the layers of cornea, epithelium, stroma or endothelium are incubated in vitro with labeled AA as precursors for the synthesis of prostaglandins and lipoxygenase metabolites.

In vivo Models

Between 4 and 6 µl of (1-^{14}C)AA as the sodium salt (1.5 µCi/eye) are injected carefully by Hamilton syringe fitted with a 36-gauge ¼-inch needle into the anterior chamber of a completely anesthetized rabbit. This approach allows for labeling of the different layers of the cornea. AA is rapidly esterified into lipids and, 4 h after the injection, less than 1% of the fatty acid remains as free fatty acid [9]. This suggests that the cornea contains very active acyltransferases that take AA quickly from the free fatty acid pool. The distribution of radioactivity changes as a function of time. Four hours after introduction of the precursor, between 40 and 75% of the fatty acid is esterified into triglycerides. Longer durations of in vivo labeling, such as 18 h, produce a redistribution of the labeled AA. By this time, 90% of the fatty acid is acylated into phospholipids [10].

An injury therefore can be applied to the labeled cornea and the metabolism of eicosanoids can be followed in corneal layers, aqueous humor, and other parts of the eye [9, 10].

Types of Lesions Applied to the Cornea

There are a variety of lesions applied to the cornea and used by researchers in order to study inflammation and the wound-healing process. The following are the ones that we use most frequently in our laboratory.

Cryogenic Lesion. A brass probe with a 3-mm diameter tip is cooled in liquid nitrogen. It is applied to the central anesthetized cornea twice, for 12 s each time, until a cryogenic 'cone' is formed in the anterior chamber. In this way the entire cornea, including the endothelium, can be wounded [11]. Neomycin ophthalmic drops are applied daily to avoid infection until reepithelialization is complete.

Endotoxin. Five microliters of 0.1% or 0.01% saline solution of purified *Escherichia coli* lipopolysaccharide B are injected under microscope into the midstroma at the center of the anesthetized cornea using a 10-µl Hamilton syringe fitted with a 33-gauge needle. This model elicits the appearance of mononuclear cells at the limbus 4 h after endotoxin injection [12].

Herpetic Keratitis Lesion. Rabbits are bilaterally injected with McKrae strain HSV (10^7 PFU/ml). One drop of the solution is placed on the minimally traumatized corneal epithelium with a 27-gauge needle. The eyelids are closed and rubbed gently against the eye surface for a few seconds. The animals are examined daily. Three days after inoculation, corneal disease can be observed under slitlamp biomicroscopy. The first manifestation of the disease is epithelial ulceration [13].

Alkali-Burn Model. A 6-mm circle of filter paper soaked in 1 *N* NaOH is applied to the anesthetized cornea for 60 s and then the eye is washed extensively with saline solution. A drop of proparacaine is applied. This lesion produces a marked burn in the cornea with irregular borders, and injures the cornea throughout the endothelium.

Inflammatory Lipid Mediators in the Cornea

Cyclooxygenase Metabolites

The release of prostaglandins in the eye after ocular irritation was first observed 20 years ago [14]. Other studies showed the presence of cyclooxygenase products in the entire cornea [15, 16] and in cell culture [8]. We monitored the changes in cyclooxygenase products (prostaglandins and thromboxane A_2) in the different layers of the cornea during the inflammatory response of corneal wound healing following in vivo and in vitro models [9, 11, 17].

Two hours after cryogenic injury, the isolated stromal and epithelial layers of the cornea incubated in vitro with 1-^{14}C-AA produce increases in total cyclooxygenase products (fig. 1). In epithelium, at 5 days after injury, there is a decline in the production of cyclooxygenase products, which reaches control values at day 15 after the lesion. In contrast, stroma and endothelium are slower than epithelium to respond by the formation of cyclooxygenase products, but reach a maximum 5 days after injury and remain elevated at 15 days [11]. The composition of cyclooxygenase products also differs among the three layers of the cornea: the major metabolites in the corneal epithelium are $PGF_{2\alpha}$ and PGE_2, in the stroma, PGI_2 and TXA_2, and in endothelium, $PGF_{2\alpha}$, PGE_2 and PGD_2, suggesting that each cell type in the cornea responds to insult in a specific manner. Most of the corneal stroma is composed of collagen fibers in which the cells, keratocytes, comprise only 2–3% of the total tissue. Despite this relatively small number of cells, the stroma seems to have a very active metabolism favoring the formation of prostaglandins and thromboxane.

The prostaglandins represent one of the most important groups of naturally occurring compounds implicated in acute inflammatory pro-

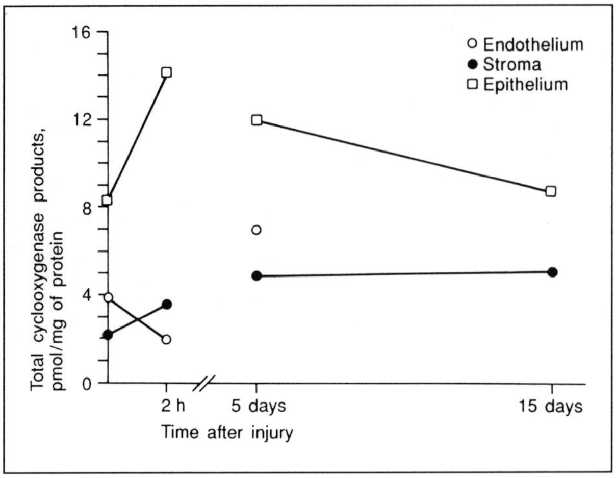

Fig. 1. Cyclooxygenase metabolites (prostaglandins and thromboxane A_2) formed in layers of the rabbit cornea during wound healing after a cryogenic injury. The values correspond to pmol/mg of protein using 1-^{14}C-AA as precursor and are calculated from reference 11.

cesses. Herpes virus infection also produces an increased formation of prostaglandins. In entire cornea the main compounds formed are PGI_2, TXA_2 and $PGF_{2\alpha}$ [13, 18], indicating a viral stimulation of the prostacyclin and thromboxane synthetases which produce these vasoactive compounds. $PGF_{2\alpha}$ is an ocular hypotensive agent [19–21], which, when applied topically to the cornea after a lesion, inhibits the synthesis of PGE_2 in the aqueous humor. $PGF_{2\alpha}$ also inhibits the increase in protein synthesis, indicating a regulatory function of this prostaglandin [22]. Studies using the in vivo model detailed above show that the major prostaglandins formed in the cornea after a cryogenic injury are $PGF_{2\alpha}$ and PGE_2, which in this model are released into the aqueous humor. Among the individual layers the stroma seems to be the corneal layer that produces the greatest release of cyclooxygenase metabolites, considering the amount of protein present in the tissue (7 mg of protein/rabbit cornea). This particular corneal layer, in vivo, also produces a great increase in PGD_2. PGD_2 is very active in mast cells, and macrophages in skin tissue are important sites of PGD_2 synthesis [23]. In addition to regulating inflammation, PGD_2 shows potent cytotoxic activity in the skin [23]. The role of these eicosanoids in cornea has not been extensively studied and deserves future investigation.

Specific binding sites have been described for most of the prostaglandins [24]. The most studied has been PGE_2, which stimulates adenylate cyclase in most tissues. In the cornea, cAMP production is stimulated by adrenergic agents [25]. Cholera toxin, which stimulates cAMP synthesis, increases the rate of wound closure in rabbit corneal epithelium [26]. PGE_2 is necessary for the maintenance of the hexagonal shape of corneal endothelial cells in culture [27]. Indomethacin, which blocks the formation of PGE_2, produces a stellate elongation of the cells. Because the cells maintain their shape in the presence of cholera toxin (which increases cAMP synthesis), an involvement of cAMP in the maintenance of corneal endothelial morphology is suggested [28]. On the other hand, in corneal endothelial cell culture, EGF mimics the effect of cell elongation during wound closure, probably acting through other receptors that stimulate the hydrolysis of polyphosphoinositols and the accumulation of inositol triphosphate [29].

Lipoxygenase Metabolites

The presence of lipoxygenase activity in the cornea has been more recently reported than cyclooxygenase activity [11, 17, 30]. 12-HETE and LTB_4 are detected in the stroma, but not in corneal endothelium, under resting conditions [17]. Cryogenic injury produces a rapid increase in the synthesis of 5-HETE and 12-HETE as early as 2 h after the lesion. The greatest increase in lipoxygenase activity is in epithelium, with a 700% increase of metabolites over control levels, as measured by HPLC [17]. Five days after the wound, levels of lipoxygenase metabolites of AA are still elevated, and are higher than cyclooxygenase products in the three layers of the cornea (fig. 2). Corneal stroma and endothelial cells have a relatively higher proportion of AA esterified to phospholipids than epithelial cells [31]. Lipoxygenase activity may reflect the synthetic capability of corneal resident cells as well as PMNs and macrophages that enter the stroma after injury. Experimental herpes simplex keratitis also produces a greater increase in lipoxygenase metabolites than in prostaglandins: LTB_4, 12-HETE, 15-HETE and 5-HETE show increases after corneal infection [13]. This increase is related to the severity of the infection [18]. Because the lipoxygenase activity is more pronounced in the central cornea than in the scleral rim, the effect of HSV infection is not occurring in the vasculature. Using the in vivo model there is an increase in 12-HETE, 5-HETE and di-HETEs (which include LTB_4) after cryogenic injury [9]. These lipoxygenase products also appear in the aqueous humor. LTB_4 is a potent

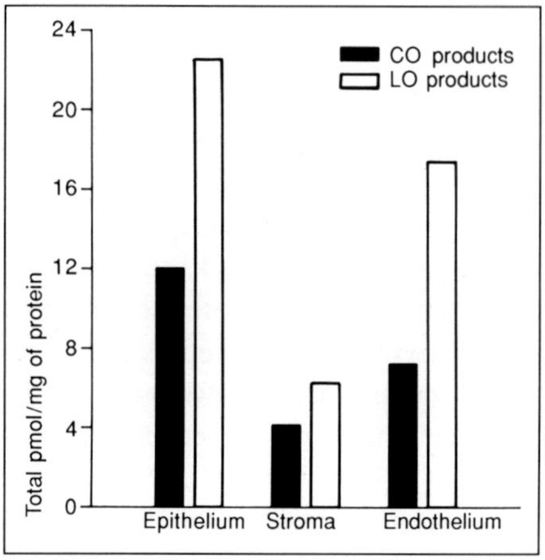

Fig. 2. Comparison between lipoxygenase and cyclooxygenase metabolites produced in corneal epithelium, stroma and endothelium 5 days after a cryogenic injury. Data calculated from reference 11. CO = Cyclooxygenase; LO = lipoxygenase.

chemotactic and aggregating agent [32]: injection of LTB_4 into the anterior chamber of rabbits causes an accumulation of PMNs in the aqueous humor [33].

Platelet-Activating Factor

Platelet-activating factor (PAF) is an ether-linked phospholipid that is produced by the combined action of a phospholipase A_2 acting on alkyl-arachidonoyl-glycerophosphocholine and an acetyltransferase, which transforms the biologically inactive lyso-PAF into the potent PAF (fig. 3). PAF is a proinflammatory agent that occurs in a variety of inflammatory cells, such as basophils, neutrophils and macrophages [34].

The presence of PAF has been demonstrated in other tissues, including the eye. Chick retina synthesizes PAF in response to neurotransmitters such as dopamine [35].

This important mediator of inflammation has been recently detected in the rabbit cornea after alkali-burn injury [36]. This injury produces a rapid synthesis of the lipid mediator that can be detected as soon as 30 min after the burn. PAF is still present 24 h after the alkali burn [37]. In these experiments, PAF is isolated by HPLC and its biological activity is deter-

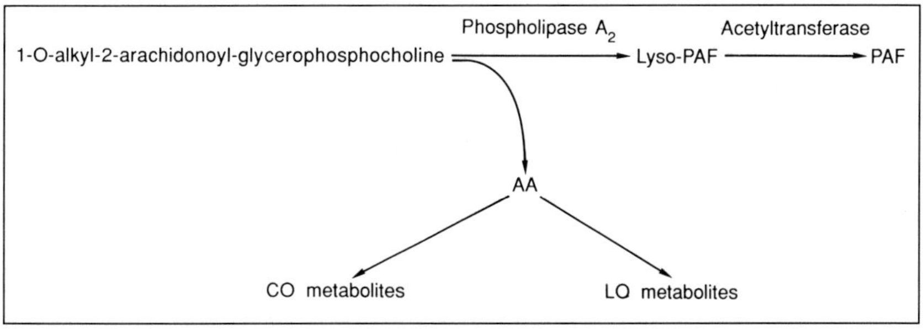

Fig. 3. Scheme of the synthesis of PAF and its relation to AA metabolites. CO = Cyclooxygenase; LO = lipoxygenase.

mined by its ability to produce platelet aggregation. This aggregation is dose-dependent and is inhibited by PAF antagonists [36]. Rabbit corneas labeled with ^{3}H-acetate in vitro also synthesize ^{3}H-PAF when stimulated by ionophore A23187, indicating the presence of acetyltransferase and a calcium-sensitive phospholipase A_2 (fig. 3), which are involved in the accumulation of PAF in the corneal response to injury or trauma [37].

Intradermal administration of PAF causes increased vascular permeability and leukocyte infiltration [38]. There is evidence that PAF stimulates a respiratory burst and chemotaxis in neutrophils [39] and macrophages [40]. PAF also produces the activation of PMNs in the cornea measured by in vitro chemiluminescence. PAF is also involved in cell-mediated immuno-graft rejections [41] and in vitro PAF suppresses the generation of cytotoxic lymphocytes important in the generation of the immune response [42]. One interesting characteristic of this lipid is that relatively little PAF is released to the medium after formation. This characteristic has led some investigators to propose that PAF may influence the inflammatory properties of other mediators of inflammation [34].

Effects of Lipoxygenase Inhibitors and PAF Antagonists on Inflammatory Lipid Mediators in the Cornea

The use of lipoxygenase inhibitors and PAF antagonists has shown that the inflammatory response in the cornea is due to the interplay of different lipid mediators.

Rabbit corneas with a cryogenic lesion incubated in vitro in the presence of different lipoxygenase inhibitors, TEI-8005 and TEI-8006, produce different patterns of inhibition [43]. TEI-8006 inhibits 5-HETE at 10 μM and 5- and 12-HETE at 100 μM; all cyclooxygenase products are partially inhibited by the drug. TEI-8005 at 100 μM is a selective inhibitor of 5-HETE. This inhibitor also decreases the synthesis of thromboxane A_2 while it increases the production of prostacyclin [43]. Another experimental lipoxygenase inhibitor, TEI-1338, used in the HSV model, is more potent, and at 10 μM inhibits 5-lipoxygenase activity, measured by its products 5-HETE and LTB_4, while at 100 μM also inhibits 12- and 15-HETE [13]. Prostaglandins are not affected. The studies with experimental lipoxygenase inhibitors demonstrate that corneal 5- and 12-lipoxygenases have different sensitivities. In addition, the use of inhibitors that regulate specific eicosanoid synthesis instead of the broader branches of the metabolism of AA can be of therapeutic interest in the control of corneal inflammation. For example, it is known that in other tissues, products like HETEs and hydroperoxyeicosatetraenoic acids (HPETEs) play a feedback role in the regulation of AA metabolism [44].

Several studies have suggested that some prostaglandins such as $PGF_{2\alpha}$ have beneficial hypotensive effects in the eye [19, 21]. As mentioned, selective lipoxygenase inhibitors such as TEI-8005 do not affect the production of $PGF_{2\alpha}$ while preferentially decreasing the synthesis of thromboxane A_2, a potent vasoconstrictor.

TEI-1338 applied topically to rabbit HSV-infected corneas produces a marked reduction in the protein content in the aqueous humor [13]. Many of the 5-lipoxygenase products increase vascular permeability: injections of LTB_4 into human skin produces local erythema and extravasation of proteins [45]. In guinea pigs, 5-lipoxygenase inhibitors reduce the permeability response of the conjunctiva [46]. Therefore, the decrease in aqueous protein caused by TEI-1338 can be attributed to its inhibition of lipoxygenase products formed in the cornea after HSV infection.

In the in vivo model of alkali-burn injury, a specific antagonist of PAF, BN-52021, applied topically to the eye, reduces the production of the lipoxygenase metabolites 5- and 12-HETEs in the cornea, while not affecting the fate of prostaglandins. PAF stimulates the synthesis of LTB_4 in neutrophils [47]. The possibility of a relationship between PAF synthesis and the lipoxygenation of AA released from the precursor of PAF in the cornea is suggested by these experiments (fig. 3).

Conclusions

The cornea responds to injury by releasing a number of lipid mediators of inflammation. Lipid response varies with the stimulus, the time after injury, and the different cell types that constitute the layers of the cornea. The use of specific inhibitors of certain pathways of AA metabolism and PAF antagonists has demonstrated that these reactions have distinct sensitivity to inhibition, and most importantly, that concerted mechanisms can modulate the inflammatory response. No doubt a better understanding of the molecular level of corneal response to inflammation will lead to new and more effective drug treatments.

References

1 Pfiester RR: The healings of corneal epithelium abrasions in the rabbit: A scanning electron microscope study. Invest Ophthalmol 1975;14:648–661.
2 Pfiester RR, Friend J, Dohlman CH: The anterior segments of rabbits after alkali burns: A metabolic and histological alteration. Arch Ophthalmol 1971;86:189–193.
3 Chusid MJ, Nelson DB, Meyer LA: The role of the polymorphonuclear leukocyte in the induction of corneal edema. Invest Ophthalmol Vis Sci 1986;27:1466–1469.
4 Meyers-Elliot RH, Chitjian PA, Billups CL: Effect of cyclosporine A on the corneal inflammatory response in herpes simplex virus keratitis. Exp Eye Res 1987;45:281–303.
5 Crosson CE, Klyce SD, Beuerman RW: Epithelial wound closure in the rabbit cornea. Invest Ophthalmol Vis Sci 1986;27:464–473.
6 Suda T, Nishida T, Ohashi Y, et al: Fibronectin appears at the site of corneal stroma wound in rabbits. Curr Eye Res 1981/1982;1:553–556.
7 Ding M, Burstein NL: Fibronectin in corneal wound healing. J Ocul Pharmacol 1988;4:75–91.
8 Taylor L, Menconi M, Leibowitz HM, et al: The effect of ascorbate, hydroperoxides and bradykinin on prostaglandins production by corneal and lens cells. Invest Ophthalmol Vis Sci 1982;23:378.
9 Bazan HEP: Corneal injury alters eicosanoid formation in the rabbit anterior segment in vivo. Invest Ophthalmol Vis Sci 1987;28:314–319.
10 Bazan HEP, Braquet P, Reddy STK, et al: Inhibition of the alkali burn-induced lipoxygenation of arachidonic acid in the rabbit cornea in vivo by a platelet-activating factor antagonist. J Ocul Pharmacol 1987;3:357–365.
11 Bazan HEP, Birkle DL, Beuerman R, et al: Cryogenic lesion alters the metabolism of arachidonic acid in rabbit cornea layers. Invest Ophthalmol Vis Sci 1985;26:474–480.
12 Howes EL, Cruse VK, Kwok MT: Mononuclear cells in the corneal responses to endotoxin. Invest Ophthalmol Vis Sci 1982;22:494–501.

13 Limberg MB, Birkle DL, Bazan HEP, et al: The effect of a new lipoxygenase inhibitor on the production of arachidonic acid metabolites during experimental herpes simplex keratitis. Curr Eye Res 1988;7:1131–1135.
14 Ambache N, Brummer MC: A simple chemical procedure to distinguish E from F prostaglandins with application to tissue extract. Br J Pharmacol Chemother 1968; 33:162–170.
15 Bhattacherjee P, Eakins KE: Inhibition of the prostaglandin synthetase systems in ocular tissues by indomethacin. Br J Pharmacol 1974;50:227–230.
16 Kass MA, Holmber NJ: Prostaglandins and thromboxane synthesis by microsomes of rabbit ocular tissue. Invest Ophthalmol Vis Sci 1979;18:166–171.
17 Bazan HEP, Birkle DL, Beuerman R, et al: Inflammation-induced stimulation of the synthesis of prostaglandins and lipoxygenation-reaction products in rabbit cornea. Curr Eye Res 1985;4:175–179.
18 Birkle DL, Sanitato JJ, Kaufman HE, et al: Arachidonic acid metabolism in eicosanoids in herpes virus-infected rabbit cornea. Invest Ophthalmol Vis Sci 1986;27: 1443–1446.
19 Camras CB, Bito LZ: Reduction of intraocular pressure in normal and glaucomatous primates *(Aotus trivirgatus)* eyes by topically applied prostaglandins F_2. Curr Eye Res 1981;1:205.
20 Stern FA, Bito LZ: Comparison of the hypotensive and other ocular effects of prostaglandins E_2 and F_2 on cat and rhesus monkey eyes. Invest Ophthalmol Vis Sci 1982;22:588–598.
21 Lee PY, Podos SM, Severin C: Effects of prostaglandin F_2 on aqueous humor dynamics of rabbit, cat and monkey. Invest Ophthalmol Vis Sci 1984;25:1087–1093.
22 Hoyng PF, Verbey N, Thorig L, et al: Topical prostaglandins inhibit trauma-induced inflammation in the rabbit eye. Invest Ophthalmol Vis Sci 1986;27:1217–1225.
23 Ikai K, Imamura S: Prostaglandin D_2 in the skin. Int J Dermatol 1988;27:141–149.
24 Robertson RP: Characterization and regulation of prostaglandins and leukotriene receptors: An overview. Prostaglandins 1986;31:395–411.
25 Neufeld AH, Zawistowski KA, Page ED, et al: Influences on the density of β-adrenergic receptors in the cornea and iris-ciliary body of the retina. Invest Ophthalmol Vis Sci 1978;17:1069.
26 Jumblatt MM, Fogle JA, Neufeld AH: Cholera toxin stimulates adenosine 3′,5′-monophosphate synthesis and epithelial wound closure in the rabbit cornea. Invest Ophthalmol Vis Sci 1980;19:1321–1327.
27 Neufeld AH, Jumblatt MM, Matkin ED, et al: Maintenance of corneal endothelial cell shape by prostaglandin E_2: Effects of EGF and indomethacin. Invest Ophthalmol Vis Sci 1986;27:1437–1442.
28 Jumblatt MM, Matkin ED, Neufeld AH: Pharmacological regulation of morphology and mitosis in cultured rabbit corneal endothelium. Invest Ophthalmol Vis Sci 1988;29:586–593.
29 Woodland J, Jumblatt MM, Neufeld A, et al: Epidermal growth factor (EGF) stimulates the hydrolysis of polyphosphoinositides and the accumulation of inositol trisphosphate in cultured corneal endothelial cells. Invest Ophthalmol Vis Sci 1988; 29(suppl):112.

30 Williams RN, Delamere NA, Paterson CA: Generation of lipoxygenase products in the avascular tissues of the eye. Exp Eye Res 1985;41:733–738.
31 Bazan HEP, Bazan NG: Composition of phospholipids and free fatty acids and incorporation of labeled arachidonic acid in rabbit cornea. Comparison of epithelium, stroma, endothelium. Curr Eye Res 1984;3:1313–1319.
32 Bray MA: The pharmacology and pathophysiology of leukotriene B_4. Br Med Bull 1983;39:249–254.
33 Bhattacherjee P, Hammond B, Salmon YA, et al: Chemotactic response to some arachidonic and lipoxygenase products in the rabbit eye. Eur J Pharmacol 1981;73: 21–28.
34 Snyder F: Chemical and biochemical aspects of platelet-activating factor: A novel class of acetylated ether-linked choline-phospholipids. Med Res Rev 1985;5:107–140.
35 Bussolino F, Gremo F, Tetta C: Production of platelet-activating factor by chick retina. J Biol Chem 1986;261:16502–16508.
36 Bazan HEP, Reddy STK, Woodland JM, et al: The accumulation of platelet-activating factor in the injured cornea may be interrelated with the synthesis of lipoxygenase products. Biochem Biophys Res Commun 1987;149:915–920.
37 Bazan HEP, Reddy STK: Production of platelet-activating factor (PAF) in the cornea after stimulation. Invest Ophthalmol Vis Sci 1989;30(suppl):404.
38 Archer CB, Page CP, Morley J, et al: Inflammatory characteristics of platelet-activating factor (PAF-acether) in human skin. Br J Dermatol 1984;110:45.
39 Gay JC, Beckman JK, Zaboy KA, et al: Modulation of neutrophil oxidative responses to soluble stimuli by platelet-activating factor. Blood 1986;67:931–936.
40 Huang SJ, Monk PN, Downes CP, et al: Platelet-activating factor-induced hydrolysis of phosphatidylinositol 4,5-bis-phosphate stimulates the production of reactive oxygen intermediates in macrophages. Biochem J 1988;249:839–845.
41 Braquet P, Touqui L, Shen TY, et al: Perspectives in platelet-activating factor research. Pharmacol Rev 1987;39:97–145.
42 Gebhardt BM, Braquet P, Bazan HEP, et al: Modulation of in vitro immune reactions by platelet-activating factor and a platelet-activating factor antagonist. Immunopharmacology 1988;15:11–20.
43 Bazan HEP: Effects of lipoxygenase inhibitors on the arachidonic acid cascade in the rabbit cornea after injury. J Ocul Pharmacol 1988;4:43–49.
44 Malle E, Heis HJ, Karadi J, et al: Lipoxygenase and hydroperoxy/hydroxyeicosatetraenoic acid formation. Int J Biochem 1987;19:1013–1022.
45 Dahlen SE, Bjok I, Hedquist P, et al: Leukotrienes promote plasma leakage and leukocyte adhesion in post-capillary venules: In vivo effects with relevance to the acute inflammation response. Proc Natl Acad Sci USA 1981;78:3887–3880.
46 Garceau D, Ford-Hutchinson AW, Charleson S: 5-Lipoxygenase inhibitors and allergic conjunctivitis reactions in the guinea pig. Eur J Pharmacol 1987;143:1–7.
47 Lin AH, Morton D, Gorman RR: Acetyl glycerol ether phosphorylcholine stimulates leukotriene B_4 synthesis in human polymorphonuclear leukocytes. J Clin Invest 1982;70:1058–1063.

Haydee E.P. Bazan, PhD, LSU Eye Center, 2020 Gravier Street, Suite B,
New Orleans, LA 70112–2234 (USA)

Bazan NG (ed): Lipid Mediators in Eye Inflammation.
New Trends Lipid Mediators Res. Basel, Karger, 1990, vol 5, pp 12–28

Eicosanoids in Ocular Inflammation

Prasad Kulkarni

Department of Ophthalmology and Visual Sciences, University of Louisville, School of Medicine, Kentucky Lions Eye Research Institute, Louisville, Ky., USA

Ocular tissues were among the first tissues from which prostaglandins (PGs) were extracted [1] and identified as PGE_2 and $PGF_{2\alpha}$ [2]. This extract from bovine iris, named 'irin', contained a mixture of PGE_2 and $PGF_{2\alpha}$ [3, 4]. Intracameral injection of irin produced miosis, raised intraocular pressure (IOP) and caused breakdown of the blood-aqueous barrier and caused conjunctival and iridial hyperemia in rabbits [1]. Although at the time it was demonstrated that the extract consisted only of PGE_2 and $PGF_{2\alpha}$, it most likely also contained other compounds that caused irritation to the rabbit eye.

Arachidonic acid (AA), the precursor of PGE_2 and $PGF_{2\alpha}$, is also converted into other biologically active PGs, i.e., PGI_2 and PGD_2 and thromboxane (Tx) A_2 by the cyclooxygenase pathway [5, 6]. Additionally, AA is converted into biologically active lipoxygenase products such as hydroxyeicosatetraenoic acid (12-, 15-HETE), 15-HETE, leukotriene (LT) B_4 and peptidoleukotrienes (LTC_4, LTD_4, LTE_4 and LTF_4) via various lipoxygenase pathways [7]. LTB_4 is chemotactic, chemokinetic and proaggregatory for polymorphonuclear leukocytes (PMN) while LTC_4, LTD_4, LTE_4 and LTF_4 are mediators of allergic reactions and asthmatic conditions [7]. Peptidoleukotrienes also have smooth muscle-stimulating properties, especially vascular, parenchymal and intestinal smooth muscles [7].

PGs (E_2 and $F_{2\alpha}$) were recently found to be effective in lowering IOP with minimal irritative response at lower doses in primates and humans [8–10]. Extensive investigations into the role of PGs, especially E_2 and $F_{2\alpha}$, were performed in rabbit ocular inflammation at the time when AA was known to be converted into only these primary stable PGs [11, 12]. Now it is known that AA is also metabolized into other products including lipoxin A and lipoxin B [13, 14], 12(S)-HETE and 12(R)-HETE via the P-450 cytochrome NADP-NADPH enzyme pathway [15–19]. Therefore, in this chapter the possible actions and interactions of these eicosanoids in ocular inflammation are discussed.

Synthesis of Eicosanoids from Arachidonic Acid

AA, a polyunsaturated fatty acid, is the major unsaturated fatty acid component of the phospholipid pool in the cell membrane. Upon stimulation of phospholipase A_2, AA is released [5–7]. Free AA is then metabolized into: (a) cyclooxygenase products; (b) lipoxygenase products; (c) lipoxins, and (d) P-450-NAD-NADPH-dependent enzyme products: 12(R)-HETEs.

Cyclooxygenase Products

AA is metabolized first into unstable cyclic endoperoxides, PGG_2 and PGH_2 (half-life about 10 min in aqueous medium). These cyclic endoperoxides are further converted into TxA_2, PGE_2, $PGF_{2\alpha}$, PGI_2 and PGD_2 [5, 6].

Lipoxygenase Products

AA is metabolized into lipoxygenase products by various lipoxygenases. For example, in platelets AA is mainly converted into 12(S)-HETE via 12-lipoxygenase while in leukocytes AA is first converted into an unstable product, 5-HETE, via the 5-lipoxygenase pathway and then into LTA_4. LTA_4 is an unstable compound and is further metabolized into a potent chemotactic product LTB_4. In tissues, such as the lung and kidney, LTA_4 is further converted into the peptidoleukotrienes, LTC_4, LTD_4, LTE_4 and LTF_4, which is a mixture of slow-reacting substance of anaphylaxis (SRS-A). These peptidoleukotrienes are thought to be mediators of asthmatic and allergic conditions [7].

Lipoxins

AA is also converted into other trihydroxy acids with a conjugated tetraene called lipoxin A and lipoxin B in leukocytes. Their biological activities are similar to that of LTs; however, these activities are considerably less potent than LTs [13, 14].

Synthesis and Release of Arachidonic Acid Metabolites in Ocular Tissues

Almost all ocular tissues synthesize cyclooxygenase and lipoxygenase products from AA in varying amounts. Also, there are tissue and species differences, for example, under identical experimental conditions the anterior uvea of the rabbit and cat synthesize about 40–45% of the total amount of cyclooxygenase products while human anterior uvea synthesizes 20–25% and bovine anterior uveal tissue produces only 5–10% of the total products (table 1).

Conjunctiva

Conjunctiva has an ability to synthesize both cyclooxygenase and lipoxygenase products from AA in substantial amounts. Rabbit conjunctiva synthesizes about 40–45% of the total cyclooxygenase products, the predominant product being PGE_2 [20–22]. Conjunctival tissue from the cat and monkey also synthesize these products [23, 24] (table 1). The predominant lipoxygenase product formed by the conjunctiva of rabbits, cats and monkeys is 12-HETE via the 12-lipoxygenase pathway [22, 25–30]. In studies using a thin-layer chromatography (TLC) technique, these tissues also synthesized LTB_4 from radiolabelled AA [22, 25, 29] (table 2). However, the synthesis of peptidoleukotrienes (LTC_4, LTD_4 and LTE_4) could not be detected in these studies because these LTs do not separate from the

Table 1. [^{14}C]-AA metabolism in various ocular tissues via the cyclooxygenase pathway

Species	Total cyclooxygenase products, %		
	conjunctiva	anterior uvea	lids
Albino rabbit	47 ± 1.0	35.1 ± 0.7	ND
Dog (adult)	5.3 ± 0.3	13.2 ± 1.2	ND
Dog (puppy)	7.5 ± 0.2	11.4 ± 0.4	ND
Cat	13.7 ± 0.2	48.5 ± 3.0	ND
Cow	ND	3.6 ± 0.2	ND
Human	ND	26.9 ± 0.5	ND
Ferret	10.8 ± 0.2	ND	ND
Cynomolgus monkey	24.3 ± 4.2	2.1 ± 0.7	27.3 ± 3.4
Rhesus monkey	19.5 ± 2.7	ND	23.1 ± 5.7

origin on the TLC plate in the solvent mixture containing ethyl acetate, isooctane, water and acetic acid. However, using the guinea pig ileum bioassay technique we demonstrated that these ocular tissues also have the ability to synthesize a slow-reacting substance (SRS) [29, 30] which represents a mixture of LTC_4, LTD_4 and LTE_4 from AA [7]. Further confirmation that SRS-like activity produced by rabbit and human conjunctiva was due to LTC_4 or LTE_4 came from studies using specific RIA techniques [29]. Interestingly, the cyclooxygenase activity was significantly higher than lipoxygenase activity in these tissues.

Anterior Uvea

Anterior uveal tissue, consisting of the iris, ciliary processes and trabecular meshwork, is highly vascularized and synthesizes substantial amounts of cyclooxygenase and lipoxygenase products. Rabbit and cat anterior uvea synthesize significantly higher amounts of cyclooxygenase products than the human anterior uvea, under identical experimental conditions [22] (table 1). The anterior uvea also synthesizes lipoxygenase products with 12-HETE being the predominant product (table 2) [26, 29]. The anterior uvea of the rabbit, cat, monkey and human, stimulated by calcium ionophore (A23187), synthesizes LTB_4 and SRS-like activity from exogenous AA [23, 24, 26, 31]. Among different species examined, the bovine iris synthesizes significantly smaller amounts of cyclooxygenase products (PGI_2 being the major product), while the formation of lipoxy-

Table 2. Arachidonate lipoxygenase products in ocular tissues[1]

Species	Tissue	5,12-di-HETE	5-HETE	12-HETE
Dutch rabbit	conjunctiva	0.75 ± 0.2	0.38 ± 0.1	2.3 ± 0.3
	iris	1.43 ± 0.2	0.31 ± 0.2	2.2 ± 0.2
Albino rabbit	conjunctiva	1.0 ± 0.3	0.5 ± 0.09	2.2 ± 0.3
Cynomolgus monkey	conjunctiva	1.54 ± 0.3	1.0 ± 0.2	2.2 ± 0.5
	iris	0.4 ± 0.1	0.4 ± 0.1	1.5 ± 0.4
	eyelid	1.8 ± 0.4	0.7 ± 0.2	2.5 ± 0.6
Rhesus monkey	conjunctiva	2.7 ± 0.6	1.6 ± 0.2	6.06 ± 2.6
	eyelid	1.12 ± 0.3	0.6 ± 0.1	2.2 ± 0.3
Human	iris	0.8 ± 0.2	0.4 ± 0.05	1.0 ± 0.2

[1] Data are mean ± SEM; product is percent of total extracted ^{14}C radioactivity.

Table 3. AA metabolism in the human retina[1]

AA metabolite	n	Percent synthesis from ^{14}C-AA
6-Keto-$PGF_{1\alpha}$	6	1.47 ± 0.73
$PGF_{2\alpha}$	4	1.75 ± 0.40
PGE_2	6	1.43 ± 0.57
PGD_2	6	0.87 ± 0.25
5,12-di-HETE	6	1.54 ± 0.52
5-HETE	4	1.40 ± 0.70
12-HETE	4	2.75 ± 1.20

n = Number of pooled samples, each containing tissue from three eyes.
[1] Data are mean ± SEM; product is percent of total ^{14}C radioactivity.

genase products was not significantly different from cat, rabbit, monkey or human iris [23] (table 1). Significance of this species difference is yet to be elucidated.

Recently, cultured human trabecular meshwork endothelial cells have been demonstrated to synthesize cyclooxygenase products, mainly PGE_2 and $PGF_{2\alpha}$ [32, 33]. Since it has been demonstrated that PGE_2 and $PGF_{2\alpha}$ decrease IOP in rabbits, cats, monkeys and humans [8–10], the endogenous PG synthesis in trabecular meshwork may be important in modulating IOP.

Other Ocular Tissues

Other ocular tissues such as the retina, lens and cornea also have the ability to synthesize cyclooxygenase and lipoxygenase products [see Bazan and Bazan, this volume]. However, these tissues have considerably less enzymatic activities than the rabbit conjunctiva and anterior uvea. Recently, we studied AA metabolism in the human retina in which cyclooxygenase activity was found to be less than that in human anterior uvea (table 3). The retinal phospholipids, in addition to AA, contain docosahexaenoic acid which is an endogenous inhibitor of cyclooxygenase [34], and this probably accounts for the decreased cyclooxygenase activity in the retinal tissues.

Recently, it was demonstrated that rabbit and bovine corneal epithelium converts AA into 12*(R)*-5,8,10,14-hydroxyeicosatetraenoic acid (compound C) and 12*(R)*-5,8-14-hydroxyeicosatetraenoic acid (compound

D) via the P-450 cytochrome NADP-NADPH-dependent enzyme [15–19]. These two compounds have biological activities and recently it was demonstrated that compound D and its synthetic isomer D_R are proinflammatory. Intracameral injection of these compounds in low doses (10 ng) caused an increase in protein concentration of the aqueous humor, chemotaxis and lowered IOP of the rabbit eye [16–19]. Also, compounds D and D_R caused neovascularization of the cornea when pellets containing these compounds were inserted into the rabbit corneal stroma. These biological activities were related to stereochemistry since other synthetic isomer 12*(S)*-HETE-*(Ds)* did not produce these biological activities in the rabbit eye [18, 19]. 12*(R)*-5,8,10,14-HETE also has biological activities including lowering IOP and inhibiting Na^+-K^+-ATPase activity of the rabbit kidney, rabbit and bovine cornea and of the rabbit iris-ciliary body [18, 19, 35].

Pathological Aspects of Eicosanoids

Cyclooxygenase Products

PGs are vasodilators and thus cause conjunctival and iridial hyperemia following topical application to the rabbit eye. In rabbits, either topical, intravitreal or intracameral administration of PGs, with the exception of 6-keto-$PGF_{1\alpha}$, increases IOP and causes the breakdown of the blood-aqueous barrier. None of the cyclooxygenase products cause PMN chemotaxis or miosis in rabbits [12, 23, 36]. However, topically applied PGE_2 and $PGF_{2\alpha}$ produced miosis in cats [9, 11].

Increased release of PGs (E_2) into the aqueous humor has been found to be associated with the collapse of the blood-aqueous barrier [12, 23, 36]. PGE_2-like activity was found in substantial amounts in the aqueous humor of patients with acute anterior uveitis, glaucomatocyclitis crisis and Behçet's disease. Such aqueous humor PGE_2-like activity was negligible in patients treated with anti-inflammatory corticosteroids [12]. Since exogenous PGs disrupt the blood-aqueous barrier and cause conjunctival and iridial hyperemia and miosis in some cases (cats), it is plausible that AA metabolites are mediators of certain types of ocular inflammation. However, the major evidence to indicate that PGs are proinflammatory came from the studies performed using the rabbit eye. In higher species such as monkey and man, lower topical doses of PGE_2 and $PGF_{2\alpha}$ reduce IOP, causing minimal or no ocular inflammation [8–11]. However, prolonged topical treatment with $PGF_{2\alpha}$ causes conjunctival hyperemia and increased corneal tissue tenderness in man [8, 37]. These PGs did not cause miosis in monkey and man [8, 11]. Thus, it appears that PGs not only have inflammatory effects but they also lower IOP depending on the dose and species.

Lipoxygenase Products

Presence of the lipoxygenase products peptidoleukotrienes LTC_4 and LTD_4 in human tear fluid following antigen challenge has been demonstrated [38]. However, topical application of LTC_4 to the human eye did not produce allergic response [39]. It is not clear yet if these compounds have any role in ocular inflammatory diseases.

Eicosanoids and Experimental Ocular Models

Experimental Ocular Inflammatory Models

Paracentesis. In the paracentesis model, the breakdown on the blood-aqueous barrier occurs within a few minutes and reaches its peak within 30–60 min and lasts up to 4–6 h without PMN infiltration.

Uveitis Induced by Endotoxin and Antigen. Endotoxin-induced uveitis is produced by intravitreal or subplantor administration of *Escherichia coli* endotoxin. The ocular inflammatory signs in this model consist of conjunctival and iridial hyperemia, breakdown of the blood-aqueous barrier and PMN infiltration into the aqueous humor and iris-ciliary body [12, 23, 36]. Similar ocular inflammatory responses can be induced by immunization procedure. Rabbits are first injected subcutaneously with antigen and later, on the 8th day, intravitreally [40]. A severe to moderate ocular inflammation can be achieved depending on the type and dose of antigen. For example, rabbit-canine sera produced severe intraocular inflammation while equine or bovine sera produced mild to moderate inflammation of the rabbit eye [40]. A modified method of Baldwin and Borgmann [40] was used by Kulkarni et al. [41] in which these workers injected bovine serum albumin (BSA) intravitreally first and then 10–15 days later BSA was injected again intravenously. In this immunogenic uveitis model and endotoxin-induced uveitis model, the ocular responses occurred sequentially and they were as follows: (1) conjunctival and iridial hyperemia; (2) collapse of the blood-aqueous barrier, and (3) PMN infiltration into aqueous humor. The peak responses from the breakdown of the blood-aqueous bar-

Table 4. Leukocyte and LTB_4 concentrations in the aqueous humor of inflamed rabbit eyes[1]

Ocular inflammatory model	PMN/mm^3	LTB_4, ng/ml
BSA (1 mg)-induced uveitis[2]	4,056 ± 597 (14)	1.3 ± 0.39 (11)
Paracentesis[3]	2 ± 2 (38)	1.5 ± 0.65 (11)

Sample size in parentheses.
[1] Data are mean ± SEM.
[2] 48 h following BSA intravitreal injection (immunogenic).
[3] 30 min after first paracentesis.

rier and PMN infiltration into aqueous humor following endotoxin administration occurred at different intervals. The peak response from the breakdown of the blood-aqueous barrier following endotoxin injection occurs within 6 h while the PMN infiltration response starts at 6 h and peaks at 16–24 h then subsides within the next 74 h.

Eicosanoids. We measured the levels of prostanoids and LTs (especially LTB_4) at various intervals during ocular inflammation in these three models using RIA techniques. Also, tissue PMN chemotaxis was measured by determining myeloperoxidase (MPO) activity in the inflamed tissue as an index of leukocyte response. The PGE_2 levels in the aqueous humor were significantly increased at the peak of the breakdown of blood-aqueous barrier response in all the models. Since exogenously applied PGs cause disruption of the blood-aqueous barrier it is possible that PGs may be responsible for this ocular inflammatory response. LTB_4 levels in the aqueous humor were also higher at a time when PMN accumulation was maximal. However, there was no quantitative relationship observed between the number of PMNs present in the aqueous humor and aqueous LTB_4 concentrations. For example, aqueous humor of the BSA-induced inflamed eye contained 1.3 ng/ml LTB_4 and approximately 4,000/mm^3 PMNs (table 4) while in the paracentesis model, LTB_4 concentration of the secondary aqueous humor was 1.5 ng/ml at a time when there were no PMNs present. It is possible that LTB_4 may be metabolized into its omega isomers and thus cannot be detected by specific LTB_4 RIA technique.

As mentioned earlier, the PMN chemotaxis (in iris-ciliary body) and PMN infiltration into aqueous humor occurred at different time periods

Table 5. MPO activity and release of LTB_4 by inflamed rabbit iris-ciliary body[1]

Ocular inflammatory model	MPO, units/mg	LTB_4, pg/mg
Endotoxin (100 ng)-induced uveitis[2]	0.51 ± 0.195 (11)	13.8 ± 5.4 (7)
Paracentesis[3]	0.0007 ± 0.001 (6)	7.1 ± 1.35 (6)

Sample size in parentheses.
[1] Data are mean ± SEM.
[2] 24 h following endotoxin intravitreal injection.
[3] 30 min after first paracentesis.

following intravitreal endotoxin. Therefore, PMN infiltration into the aqueous humor may not be a true index of PMN chemotaxis in iris-ciliary processes. We measured MPO activity (an index of tissue PMN chemotaxis) and release of LTB_4 from the endotoxin-induced inflamed iris-ciliary process into Krebs-Henseleit medium following 30 min incubation at 37 °C. Interestingly, there was then an excellent correlation observed between MPO activity, i.e., PMN chemotaxis and release of LTB_4 from the iris-ciliary body (table 5), indicating that LTB_4 may play a role in tissue PMN chemotaxis response. Further evidence to indicate that this 5-lipoxygenase product plays an important role came from the studies in which a 5-lipoxygenase inhibitor BW 218C87 [42] significantly inhibited PMN response without affecting breakdown of the blood-aqueous barrier response.

In earlier studies [12, 23, 36], the role of chemical mediators in ocular inflammation was assessed by measuring the release of mediators and inflammatory responses in the aqueous humor. In the present study, we demonstrated that: (1) the peak PMN response in the aqueous humor and tissue were at different intervals, and (2) lack of quantitative correlation between aqueous PMNs and LTB_4 concentrations of the aqueous humor. These observations question the true validity of measuring PMN response and levels of chemical mediators in the aqueous humor to assess the role of chemical compounds in PMN chemotaxis.

Other lipoxygenase products such as peptidoleukotrienes are also found in rabbit aqueous humor but LTC_4 levels were not significantly different in the aqueous humor in either control or inflamed rabbit eye [unpubl. observation]. LTC_4 administered topically or intracamerally to the cat eye caused miosis without affecting the breakdown of the blood-aqueous barrier or PMN response [42], while LTD_4 increased microvascular permeability of the guinea pig conjunctiva. This response was blocked by selective peptidoleukotriene antagonists [43]. LTD_4 also caused eosinophil infiltration and potentiated the PMN chemotaxis in response to LTB_4 in guinea pig conjunctiva. Although the role of LTC_4 or LTD_4 is not known in clinical ocular allergic reactions, increased release of LTC_4 or LTD_4-like activity is demonstrated into tear fluid of patients with ocular allergic diseases such as conjunctivitis [38].

Other Mediators. The role of other mediators of inflammation in ocular inflammation such as histamine, interleukin-1, platelet-activating factor (PAF), cachectin and substance P are discussed elsewhere in this book [see Bhattacherjee and Paterson, this volume].

Anti-Inflammatory Agents

As further evidence to indicate that AA metabolites are, indeed, one of the chemical mediators of ocular inflammation, steroidal and nonsteroidal anti-inflammatory agents that interfere with AA metabolism were used [23, 41, 45]. Corticosteroids such as prednisolone and dexamethasone were effective in inhibiting external ocular inflammatory responses in the corneal wound model [23] as well as in the BSA-induced uveitis model [41]. Corticosteroids inhibit phospholipase A_2 which is responsible for the release of AA from the phospholipid pool and thus prevents AA metabolism and inhibits inflammation [23].

Nonsteroidal anti-inflammatory drugs such as aspirin or aspirin-like drugs, indomethacin, ketoprofen, flurbiprofen, diclofenac, enolicam inhibited the PMN release into tear fluid in the corneal wound model [23, 44]. However, lower doses of indomethacin, ketoprofen aspirin, in fact, potentiated the PMN response [41, 44]. Also, in the BSA-induced uveitis model indomethacin inhibited the breakdown of the blood-aqueous barrier response but potentiated the PMN infiltration into the aqueous humor [41]. Thus, it was believed that selective inhibition of PGE_2 (responsible for breakdown of the blood-aqueous barrier response) leads to significant reduction in the breakdown of the blood-aqueous barrier response; however, selective inhibition of the cyclooxygenase pathway by aspirin-like drugs may lead to the formation of more lipoxygenase products, especially LTB_4 which will potentiate the PMN response [23, 44, 46]. We tested this hypothesis in the endotoxin-induced uveitis model and observed that even up to 100% inhibition of the cyclooxygenase pathway did not facilitate formation of LTB_4 in control as well as in inflamed anterior uvea [unpubl. observation].

Bhattacherjee et al. [27, 47] have demonstrated that intracameral LTB_4 is a potent chemotactic agent for PMNs in the rabbit and rat and that high doses of intracameral PGE_2 inhibit PMN chemotaxis induced by intracameral LTB_4. Recently, we demonstrated [48, 49] that indomethacin was ineffective in inhibiting PMN response in the endotoxin-induced uveitis model. But a hydroxamic acid analogue [50] which is a potent selective 5-lipoxygenase inhibitor given orally three times daily significantly inhibited the PMN response but not the breakdown of the blood-aqueous barrier [49]. These observations indicate that LTB_4, indeed, is at least one of the mediators of PMN chemotaxis in this model. In the paracentesis model, indomethacin or aspirin inhibit the breakdown of the blood-aqueous barrier suggesting PGE_2 is involved in this response.

Eicosapentaenoic Acid Metabolism in Ocular Tissues

Eicosapentaenoic acid (EPA) is a C20:5 polyunsaturated fatty acid mainly found in the fish diet of Eskimos. Arctic sea animals such as whale, cod and mackerel are found to have fatty acids like EPA, docosahexaenoic acid and AA in their phospholipid pool membrane. Similar to AA, EPA is metabolized into cyclooxygenase (3-series) and lipoxygenase products (5-series). Patients with atherosclerosis or other cardiovascular diseases are recommended to eat EPA-enriched diet for the following reasons. Caucasian diet (red meat) contains mainly AA which is converted into PGs, PGI_2 and TxA_2. PGI_2 is a vasodilator, antiplatelet aggregatory agent and increases adenylase cyclase activity, while TxA_2 is a potent vasoconstrictor, platelet-aggregating agent and lowers cAMP in platelet [6, 51]. Thus, maintenance of the physiological hemodynamic nature depends upon the balance between the production of PGI_2 by blood vessels and TxA_2 by platelets. This balance is apparently disrupted in cardiovascular diseases since atherosclerotic blood vessels are deficient in the production of PGI_2 [52–54]. PGI_3 (an EPA product), although less potent, is also a vasodilator and antiplatelet-aggregating agent, while TxA_3 has no biological activity. Additionally, EPA is a competitive inhibitor of AA-cyclooxygenase and lipoxygenase enzymes and it lowers plasma cholesterol and low density plasma lipids (LDL) [6, 51, 55]. Thus, inhibition of AA metabolism by EPA probably lowers the incidence of cardiovascular diseases. Perhaps this is why Eskimos whose main diet consists of fish have a lower incidence of cardiovascular diseases [6, 51, 55]. It is interesting to note that Eskimos also have been reported to have a lower incidence of open angle glaucoma but no incidence of closed angle glaucoma when compared with that of Caucasians [56–59]. However, the relevance of a fish diet and the low incidence of open angle glaucoma in Eskimos is not known.

EPA is converted into the 3-series of cyclooxygenase products; PGE_3, $PGF_{3\alpha}$, 17-delta-6-keto-$PGF_{1\alpha}$ (stable product of PGI_3), PGD_3 and TxB_3 (a stable product of TxA_3) in the rabbit conjunctiva and human and rabbit anterior uvea [6, 51, 60]. Cynomolgus and rhesus monkey conjunctiva and eyelid also synthesize smaller but detectable amounts of 3-series of PGs from EPA [24]. Although cyclooxygenase products are formed from exogenous EPA, incorporation of ^{3}H-EPA in the phospholipid pool of human or rabbit anterior uvea was negligible [unpubl. observation]. PGE_3 and PGD_3 have been found to lower IOP in the rabbit eye [61]; however, since EPA is not stored in the phospholipid pool of the anterior uvea, it will be interesting to study the effect of EPA-enriched diet on IOP. It is possible that EPA, in this case, may act as an inhibitor for the AA-cyclooxygenase pathway.

EPA, similar to AA, is converted into the 5-series of lipoxygenase products. It is converted into 12-HEPE via 12-lipoxygenase while 5-lipoxygenase products of EPA are LTB_5, LTC_5 and LTD_5. EPA is a competitive inhibitor of the AA-lipoxygenase pathway. 5-Lipoxygenase products of EPA are inflammatory mediators but are less potent than AA-lipoxygenase products [62, 63]. Thus, an EPA-enriched diet in the rabbit had very little or no effect in the inflammatory responses in the endotoxin-induced uveitis model [64].

Conclusion

There is enough evidence now to indicate that eicosanoids such as cyclooxygenase products PGE_2 and $PGF_{2\alpha}$ and lipoxygenase products LTB_4, LTC_4 and LTD_4 play a role in the normal as well as pathophysiological functions of the eye, depending upon the concentrations and species.

$PGF_{2\alpha}$ lowers the IOP in higher species such as monkey and human eye. On the other hand, there is substantial evidence to suggest that break-

Table 6. Nonsteroidal anti-inflammatory drugs in clinical trials

1	Prevention of pupillary constriction during cataract surgery
2	Prevention of postoperative inflammation, i.e., disruption of blood-aqueous humor barrier and IOP following cataract surgery or lens implantation or laser trabeculoplasty
3	Prevention of neovascularization induced by either contact lens or corneal wounds
4	Improvement of lens opacity (Bendazac)
5	Prevention of cystoid macular edema following intraocular surgery; postoperative inflammation leads to increased permeability of choroidal vasculature; therefore, long-term pre- and postoperative treatment may be effective; postoperative treatment is not effective
6	Prevention of conjunctional and iridial hyperemia
7	Prevention of allergic ocular diseases

down of the blood-aqueous barrier response is due to PGE_2 while PMN chemotaxis is due to the action of LTB_4. In certain species such as the cat, PGs are miotic. Although there is not direct evidence to show that PGs are miotic in humans, flurbiprofen and indomethacin are widely used by clinicians during cataract surgery. Also, other nonsteroidal anti-inflammatory agents that interfere with AA metabolism are in clinical trials (table 6).

Acknowledgements

I would like to thank Dr. Parimal Bhattacherjee for his valuable advice while preparing the manuscript, Dr. A. Abdel-Latif for his assistance in measuring eicosapentaenoic acid uptake studies, the Kentucky Lions Eye Bank for providing human retinal tissues, Ms. Terri Trotter for the typing and editorial assistance in the preparation of the manuscript, and Mr. Gregg Shirk and Ms. Leanna Smith for technical assistance. This work was supported by the National Eye Institute, NEI Grant No. RO1 EYO2861.

References

1 Ambache N: Properties of irin, a physiological constituent of the rabbit's iris. J Physiol (Lond) 1957;129:65–66.

2 Ambache N, Brummer HC: A simple chemical procedure for distinguishing E from F prostaglandin with application to tissue extracts. Br J Pharmacol Chemother 1968; 33:162–170.

3 Bergstrom S, Rhyage J, Samelsson B, et al: The structures of prostaglandins E_1, $F_{1\alpha}$ and $F_{2\alpha}$. J Biol Chem 1986;238:3555–3558.

4 Anggard E, Samuelsson B: Smooth muscle stimulating lipids in sheep iris. The identification of prostaglandin $F_{2\alpha}$. Biochem Pharmacol 1964;13:281–283.

5 Moncada S, Vane JR: Prostacyclin in perspective; in Vane JR, Bergstrom S (eds): Prostacyclin. New York, Raven Press, 1979, pp 5–16.

6 Moncada S, Vane JR: Prostacyclin in the cardiovascular system; in Samuelsson B, Ramwell PW, Peoletti R (eds): Advances in PGs, Thromboxane and Leukotrienes. New York, Raven Press, 1980, vol 6, pp 43–60.

7 Samelsson B: Leukotrienes: A new group of biologically active compounds including SRS-A. Trends Pharmacol Sci 1980;9:227–228.

8 Bito LZ: Prostaglandins. Arch Ophthalmol 1987;105:1036–1039.

9 Bito LZ, Camaras CB, Gum GG, et al: The ocular hypotensive effects and side effects of prostaglandins on the eyes of experimental animals; in Bito LS, Stjernschantz J (eds): The Ocular Effects of Prostaglandins and Other Eicosanoids. New York, Liss, 1989, pp 349–368.

10 Bito LZ, Stjernschantz J: The ocular effects of eicosanoids and other autocoids; in Bito LS, Stjernschantz J (eds): Historic Background and the Need for a Broader Perspective. New York, Liss, 1989, pp 1–18.

11 Bito LZ, Draga, A, Blanko K, et al: Long-term maintenance of reduced intraocular pressure by daily or twice daily topical application of prostaglandins to cut on rhesus monkey eyes. Invest Ophthalmol Vis Sci 1983;24:312–319.

12 Eakins KE: Prostaglandins on the eye; in Karim SM (ed): Prostaglandins: Physiological, Pharmacological and Pathological Aspects. Baltimore, University Park Press, 1976, pp 63–81.

13 Serhan CN, Hamberg M, Samuelsson BP: Lipoxin: Novel series of biologically active compounds formed from arachidonic acid in human leukocytes. Proc Natl Acad Sci USA 1984;81:5335–5339.

14 Serhan CN, Hamberg M, Ramstedt U, et al: Lipoxins: stereochemistry, biosynthesis and biological activities; in Samuelsson B, Ramwell PW, Paoletti R (eds): Advances in Prostaglandins, Thromboxane and Leukotriene Research. New York, Raven Press, 1986, vol 16, pp 83–87.

15 Schwartzman ML, Abraham NG, Masferrer J, et al: Cytochrome P-450-dependent arachidonate metabolism in corneal epithelium: Formation of biologically active compounds; in Samuelsson B, Paoletti R, Ramwell PW (eds): Advances in Prostaglandins, Thromboxane and Leukotriene Research. New York, Raven Press, 1987, vol 17A, pp 78–83.

16 Schwartzman ML, Balazy M, Masferrer JL, et al: 12(R)-hydroxyeicosatetraenoic acid: A cytochrome P-450-dependent arachidonate metabolite that inhibits Na^+,K^+-ATPase in the cornea. Proc Natl Acad Sci USA 1987;84:8125–8129.

17 Schwartzman ML, Dunn MW, Masferrer JL: 12(R)-hydroxyeicosatetraenoic acid, a novel corneal arachidonate metabolite lowers intraocular pressure. Invest Ophthalmol Vis Sci 1988;29(suppl):85.

18 Masferrer JL, Dunn MW, Schwartzman ML: 12(R)-hydroxyeicosatetraenoic acid: Angiogenic cytochrome P-450-dependent metabolite of arachidonic acid from the corneal epithelium. Proc Int Soc Eye Res 1988;5:31.

19 Masferrer JL, Dunn MW, Schwartzman ML: Ocular effects of a novel corneal arachidonate metabolite. Invest Ophthalmol Vis Sci 1988;29(suppl):122.
20 Bhattacherjee P, Kulkarni PS, Eakins KE: Metabolism of arachidonic acid in rabbit ocular tissues. Invest Ophthalmol Vis Sci 1979;18:173–178.
21 Kass MA, Holmbert NT: Prostaglandin and thromboxane synthesis by microsomes of rabbit ocular tissues. Invest Ophthalmol Vis Sci 1979;18:166–171.
22 Kulkarni P, Srinivasan BD: Steroidal and nonsteroidal anti-inflammatory drugs in ocular inflammation. Ocul Inflam Ther 1983;1:11–18.
23 Kulkarni P, Srinivasan BD: Nonsteroidal anti-inflammatory drugs: Mechanical and clinical use; in Lewis A, Furst DE (eds): Nonsteroidal Anti-Inflammatory Drugs in Ocular Inflammatory Conditions. New York, Dekker, 1986, pp 107–155.
24 Kulkarni PS, Kaufman PL, Srinivasan DB: Eicosapentaenoic acid metabolism in cynomolugus and rhesus conjunctiva and eyelid. J Ocul Pharmacol 1987;3:349–356.
25 Williams RN, Paterson CA: Modulation of corneal lipoxygenase by ascorbic acid. Exp Eye Res 1986;43:7–13.
26 Bhattacherjee P, Kulkarni PS, Eakins KE: Differential inflammatory effects of arachidonic acid on rabbit conjunctiva and iris: A possible role of lipoxygenase in the conjunctival response; in Samuelsson B, Ramwell PM, Paoletti R (eds): Advances in Prostaglandin, Thromboxanes and Leukotrienes Research. New York, Raven Press, 1980, vol 8, pp 1727–1731.
27 Bhattacherjee PS, Hammond B, Salmon JR, et al: Effect of lipoxygenase products on leukocyte accumulation in the rabbit eye; in Samuelsson B, Paoletti R (eds): Leukotrienes and Other Lipoxygenase Products. New York, Raven Press, 1982, pp 325–330.
28 Bhattacherjee P, Eakins KE: Lipoxygenase products: Mediation of inflammatory responses and inhibition of their formation; in Chavkin LW, Bailey DM (eds): Leukotriene Chemistry and Biology. London, Academic Press, 1984, pp 195–214.
29 Kulkarni PS, Srinivasan BD: Synthesis of slow-reacting substance-like activity in rabbit conjunctiva and anterior uvea. Invest Ophthalmol Vis Sci 1983;24:1079–1085.
30 Kulkarni PS, Srinivasan BD: Human anterior uvea synthesizes lipoxygenase products from arachidonic acid. Invest Ophthalmol Vis Sci 1984;25:221–223.
31 Kulkarni PS, Kaufman PL, Srinivasan BD: Cyclooxygenase and lipoxygenase pathways in cynomolgus and rhesus monkey conjunctiva, anterior uvea and eyelids. Curr Eye Res 1987;6:801–808.
32 Weinreb RN, Mitchell MD, Polanski JR: Prostaglandin production by human trabecular cells: In vitro inhibition by dexamethasone. Invest Ophthalmol Vis Sci 1983;24:1541–1545.
33 Weinreb RN, Mitchell MD: Experimental investigations of intraocular eicosanoids: Cultured human trabecular cells and laser photocoagulations of the rabbit iris. Curr Eye Res 1985;4:281–290.
34 Lokesh BR, Black JM, German JB, et al: Docosahexaenoic acid and other dietary polyunsaturated fatty acids suppress leukotriene synthesis by mouse peritoneal macrophage lipods. Lipids 1988;23:968–972.
35 Socci R, Bhattacherjee P, Delamere N: The influence of 12(R)-HETE upon ATPase activity in the rabbit ciliary epithelium. Invest Ophthalmol Vis Sci 1988;30(suppl):447.

36 Bhattacherjee P: The role of arachidonate metabolites in ocular inflammation; in Bito LS, Stjernschantz J (eds): The Ocular Effects of Prostaglandins and Other Eicosanoids. New York, Liss, 1989, pp 211–227.
37 Alms A, Villumsen J: Effects of topically applied $PGF_{2\alpha}$ and its isopropylesters on normal and glaucomatous eyes; in Bito LZ, Stjernschantz J (eds): The Ocular Effects of Prostaglandins and Other Eicosanoids. New York, Liss, 1989, pp 437–446.
38 Bisggard H, Ford-Hutchinson AW, Charleson S: Production of peptido-leukotrienes in human tear fluid following antigen challenge. Prostaglandins 1984;28:620–628.
39 Weston JH, Abelsson MB: Leukotriene C_4 in rabbit and human eyes. Invest Ophthalmol Vis Sci 1985;26(suppl):191.
40 Baldwin HA, Borgman AR: An improved immunogenic uveitis test in rabbits for evaluation of anti-inflammatory compounds (36482). Proc Soc Exp Biol Med 1979; 133:1326–1330.
41 Kulkarni PS, Bhattacherjee PS, Eakins KE, et al: Anti-inflammatory effects of betamethasone phosphate, dexamethasone phosphate and indomethacin on rabbit ocular inflammation induced by bovine serum albumin. Curr Eye Res 1981;1:43–49.
42 Stjernschantz J, Truman S, Sears M: Ocular responses to leukotriene C_4 and D_4 in the cat. Prostaglandins 1984;27:5–15.
43 Woodward DF, Nievas AL, Gary NK, et al: Studies with a potent leukotriene-antagonist, SKF 104353, reveal that peptido-leukotrienes play a major role in mediating experimental allergic conjunctivitis. Invest Ophthalmol Vis Sci 1987; 28(suppl):200.
44 Srinivasan BD, Kulkarni PS: Inhibitors of the arachidonic acid cascade in the management of ocular inflammation; in Bito LS, Stjenuschantz J (eds): The Ocular Effects of Prostaglandins and Other Eicosanoids. New York, Liss, 1989, pp 229–250.
45 Kulkarni PS, Srinivasan BD: Effect of topical and intraperitoneal indomethacin on generation of PGE_2-like activity in rabbit conjunctiva and iris-ciliary body. Exp Eye Res 1981;33:121–129.
46 Higgs GA, Flower RS, Vane JR: A new approach to anti-inflammatory drugs. Biochem Pharmacol 1979;28:1954–1961.
47 Bhattacherjee P, Eakins KE, Hammond B: Chemotactic activity of arachidonic acid lipoxygenase products in the rabbit eye. Br J Pharmacol Chemother 1981;72:254P-255P.
48 Kulkarni PS: Role of eicosanoids in ocular inflammation. Proc Int Soc Eye Res 1988;5:85.
49 Kulkarni PS, Bhattacherjee PS: Effects of selective cyclooxygenase and lipoxygenase inhibitors on the ocular injury and inflammatory models. Invest Ophthalmol Vis Sci 1989;30(suppl):375.
50 Summers JB, Mazdiyasmi H, Homs JH, et al: Hydroxamic acid inhibitors of 5-lipoxygenase. J Med Chem 1987;30:574–580.
51 Needleman P, Sprecher H, Whitaker MO, et al: Mechanism underlying the inhibition of platelet aggregation by eicosapentaenoic acid and its metabolits; in Samuelsson B, Ramwell PW, Paoletti R (eds): Advances in Prostaglandin, Thromboxane and Leukotriene Research. New York, Liss, 1980, pp 61–68.
52 Sinzinger H, Fiegl W, Silberbaur K: Prostacyclin generation in atherosclerotic arteries. Lancet 1979;ii:469.

53 Gryglewski RJ, Dembinska-Kiec A, Zmuda A, et al: Prostacyclin and thromboxane A_2 biosynthetic capabilities of heart, arteries and platelets at various stages of experimental atherosclerosis in rabbits. Atherosclerosis 1979;31:385–394.

54 Geling NG, Neumov IS, Antonov AS: Prostacyclin formation by human aorta in different types of atherosclerotic lesion. Biomed Biochim Acta 1984;43:S281–S283.

55 Kamido H, Matsuzawa Y, Selichiro T: Lipid composition of platelets from patients with atherosclerosis: Effect of purified eicosapentaenoic acid ethyl ester administration. Lipids 1988;23:917–923.

56 Cox J: Angle-closure glaucoma among the Alaskan Eskimo. Arctic Ophthalmol 1984;6:135–137.

57 Drance SM: Angle-closure glaucoma among Caucasian Eskimos. Can J Ophthalmol 1976;8:252–254.

58 Clemmesen V, Alsbiek PH: Primary angle-closure glaucoma (a:c:g) in Greenland. Acta Ophthalmol 1971;49:47–48.

59 Albirk PH: Angle-closure glaucoma surveys in Greenland Eskimos. Can J Ophthalmol 1973;8:260–264.

60 Kulkarni PS, Srinivasan BD: Eicosapentaenoic acid metabolism in human and rabbit anterior uvea. Prostaglandins 1986;31:1159–1164.

61 Kulkarni PS, Srinivasan BD: Prostaglandin E_3 and D_3 lower intraocular pressure. Invest Ophthalmol Vis Sci 1985;26:1178–1182.

62 Terano T, Salmon JA, Moncada S: Biosynthesis and biologic activity of leukotriene B_5. Prostaglandins 1984;27:217–232.

63 Terano T, Salmon JA, Higgs GA, et al: Eicosapentaenoic acid as a modulator of inflammation: Effect on prostaglandin and leukotriene synthesis. Biochem Pharmacol 1986;35:779–785.

64 Rubin RM, Bennet WM, Elzinga L, et al: Effects of a fish oil dietary supplement of endotoxin-induced ocular inflammation. J Ocul Pharmacol 1988;4:259–267.

Prasad Kulkarni, PhD, Department of Ophthalmology and Visual Sciences, University of Louisville, School of Medicine, Kentucky Lions Eye Research Institute, Louisville, KY 40292 (USA)

Bazan NG (ed): Lipid Mediators in Eye Inflammation.
New Trends Lipid Mediators Res. Basel, Karger, 1990, vol 5, pp 29–40

Lipid Mediators and Conjunctival Inflammation

Salim I. Butrus[a], *Lisa M. Smith*[b], *Mark B. Abelson*[b]

[a]George Washington University Medical Center, Washington, D.C., and
[b]The Eye Research Institute, Harvard Medical School, Boston, Mass., USA

The conjunctiva is a thin, continuous mucous membrane which lines the posterior lid surface and extends onto the globe to cover the anterior sclera interfacing with the corneal epithelium at the limbus. It is divided into three areas: the palpebral conjunctiva refers to that on the posterior side of the lids; the bulbar conjunctiva covers the anterior sclera and the two tissues meet in the cul-de-sac to form the fornix. There are two layers to the conjunctiva: the epithelium and the substantia propria. Goblet cells, lymphocytes and mast cells are abundant in the epithelial layer and superficial substantia propria. The conjunctiva is innervated by the fifth cranial nerve. However, there are few pain fibers thus explaining why pain is not usually associated with the intense conjunctival inflammation which can be seen.

In the past 20 years, great efforts have been made to elucidate the events which occur after conjunctival injury takes place. Conjunctival inflammation is always preceded by some stimuli: either mechanical, thermal, microbial, or biochemical. Tissue injury is followed by the release and local accumulation of numerous potent mediators with specific functions, such as initiation or modulation of the inflammatory or immune response. Vasoactive compounds such as histamine and bradykinin are preformed mediators released upon injury or mast cell degranulation, thus eliciting some general effects, such as vasodilation (H_2 receptor activation), local increase in blood flow, heat production, and vessel permeability leading to edema. Other preformed mediators released from the mast cell, such as eosinophil chemotactic factor, elicit the immediate infiltration of eosinophils. The diffusion of inflammatory cells to the inflamed tissue is the next histologically evident event. However, it is the biochemical activation of the arachidonic acid (AA) cascade of prostaglandins and leukotrienes which initiates this step. Although the responsibilities of these lipid mediators in conjunctival inflammation have not been completely elucidated; the role of many specific prostaglandins and leukotrienes has been identified in some inflammatory reactions. Much information has been gained from studying the effects of exogenous application of specific lipid mediators to the ocular surface.

In this chapter, the focus is on the role of lipid mediators in conjunctival inflammation, specifically in those conditions in which the conjunctiva, and sometimes the peripheral cornea, are the main target tissues. These conditions may well be mast cell-mediated such as in ocular allergy; immune complex-mediated, such as in rheumatoid arthritis and non rheumatic vascular diseases [1–3], or mucous membrane diseases such as cicatricial pemphigoid [4].

Topical Arachidonic Acid Model

Lipid mediators released in the conjunctival tissue are the final products of a common pathway triggered by breakdown of phospholipids and release of AA by phospholipase, with subsequent conversion to active compounds by two major enzymes: cyclooxygenase and 5-lipoxygenase. The ocular effects of topical arachidonate have been well documented. They include conjunctival vasodilation, increase intraocular pressure, pupillary constriction and increased aqueous humor proteins and prostaglandins [5–7].

Topical cyclooxygenase and lipoxygenase inhibitors have proven to be beneficial in the treatment of some conditions such as vernal and giant papillary conjunctivitis [8–11]. The lack of systemic toxicity is one important benefit from the use of topical anti-inflammatory agents. An accurate and predictive model of conjunctival inflammation is needed in which different nonsteroidal anti-inflammatory agents can be screened, evaluated, and compared.

Topical administration of AA has been used for this purpose in rabbits (0.5%) and in humans (0.25%) [12]. These concentrations are the maximum doses of AA which induce only a conjunctival response without any anterior chamber reaction, miosis or elevation in intraocular pressure.

AA Rabbit Model

New Zealand albino rabbits, weighing approximately 2 kg, with no evidence of ocular inflammation, were used. Each rabbit received 20 μl of a cyclooxygenase or lipoxygenase inhibitor (aspirin, indomethacin, piroxicam (cyclooxygenase inhibitors), hydroxyarachidonamide (HA) or phenidone (lipoxygenase inhibitors)) in one eye and 20 μl of the respective vehicle in the fellow eye in a masked fashion (n = 12). Ten minutes later, 20 μl of 0.5% sodium arachidonate were instilled in both eyes.

All eyes were evaluated clinically for signs of conjunctival inflammation every 15 min for 1 h and also photographed at 1 h for documentation. Parameters of inflammation were graded on a scale of 0 to 3+. The parameters included lid closure, conjunctival chemosis, mucous discharge and hyperemia. For lid closure: 0 = eye completely open, 1+ = 1/3 closed, 2+ = 2/3 closed and 3+ = completely closed. The remainder of the parameters were graded as follows: 0 = none, 1+ = mild, 2+ = moderate, and 3+ = severe.

AA Human Model

Seven normal human volunteers who had signed an informed consent, and not used any eye or systemic medications for at least 48 h prior to the study, were included. One eye of each volunteer was pretreated topically with 20 μl of aspirin, indomethacin, prioxicam, or HA. The fellow eye received 20 μl of the respective vehicle in a double-masked fashion. Ten minutes later, both eyes received 20 μl of 0.25% sodium arachidonate. Each subject received each inhibitor after a rest period of at least 48 h between testing sessions. Slit lamp examinations were performed and symptoms and signs evaluated at baseline, 15 min, 30 min, 1 h and 2 h, to assess conjunctival hyperemia, chemosis, burning sensation, and mucous discharge. All parameters were graded on a scale of 0 to 3+. Photographs were also taken.

Results of the Rabbit Model

AA produced conjunctival inflammation in all rabbits (fig. 1). Eyes treated with the vehicle were not affected. The time course of the reaction is shown in figure 2, which shows the marked conjunctival inflammation produced by 0.50% AA at 1 h. Pretreatment with aspirin, indomethacin or piroxicam, dramatically inhibited all four parameters of conjunctival inflammation (fig. 3). Neither HA nor phenidone blocked any of the parameters (fig. 4).

Results of the Human Model

In all subjects, 0.25% AA produced mild conjunctival redness which peaked at 30 min and completely resolved in approximately 2 h. Pretreat-

1

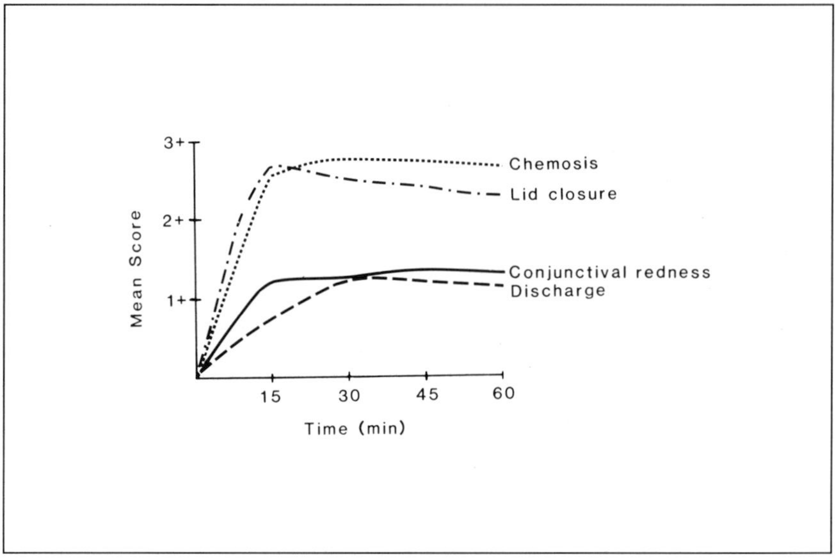

2

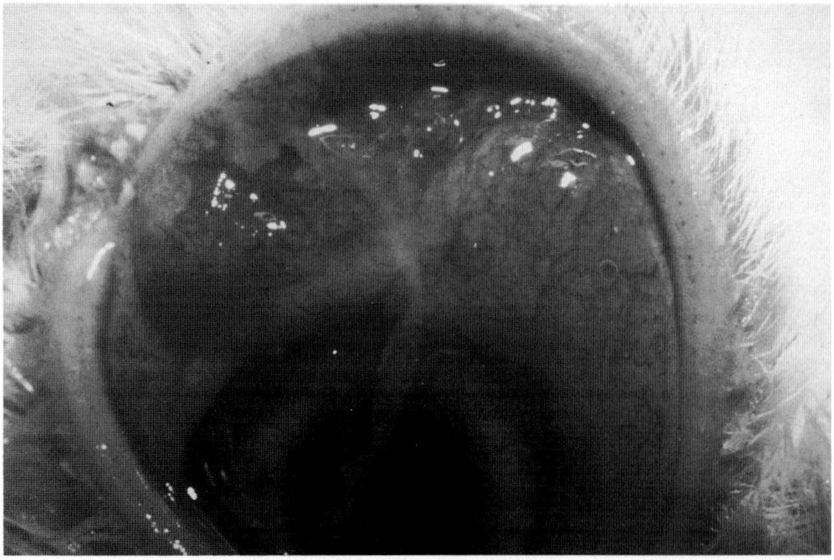

Fig. 1. Inflammatory response of rabbit eyes to topical AA [from 12, with permission].

Fig. 2. Response of rabbit eye 1 h after AA. Note marked conjunctival redness, chemosis and mucous discharge [from 12, with permission].

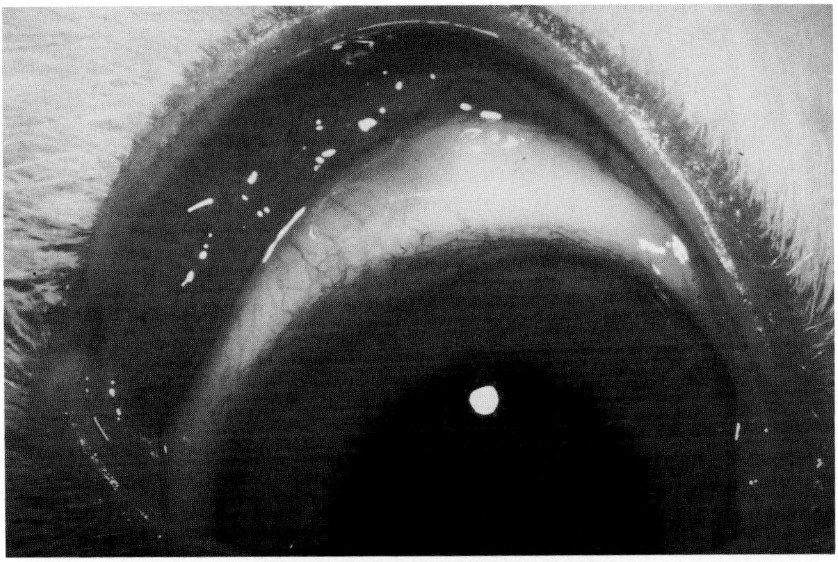
3

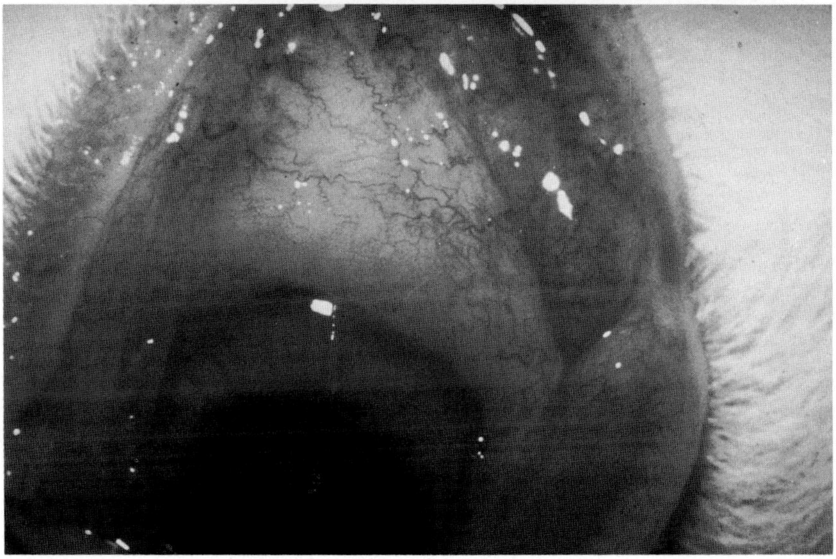
4

Fig. 3. Inhibition of conjunctival inflammation by pretreatment with aspirin, indomethacin and piroxicam [from 12, with permission].

Fig. 4. Failure of HA and phenidone to inhibit conjunctival inflammation [from 12, with permission].

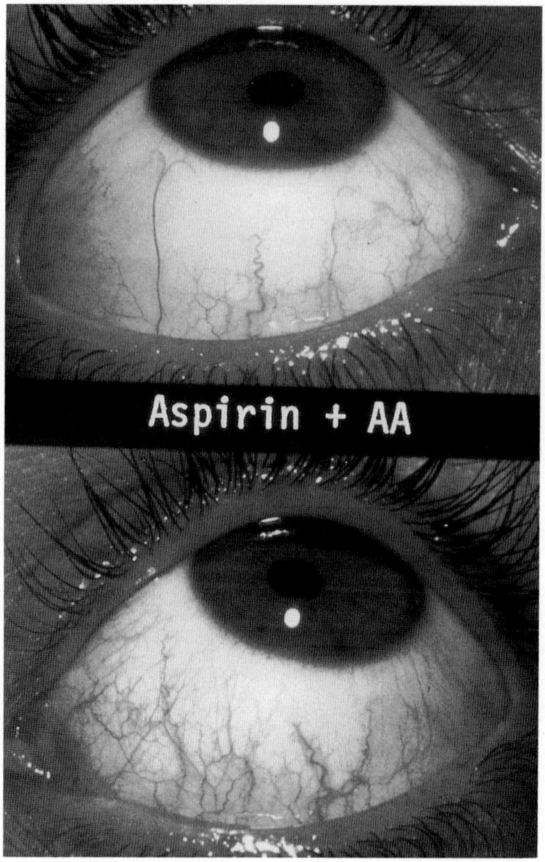

Fig. 5. Conjunctival redness induced by topical 0.25% AA in a human eye (bottom). Note reduction of redness caused by pretreatment with aspirin [from 12, with permission].

ment with aspirin or indomethacin significantly reduced conjunctival redness (fig. 5); however, piroxicam and HA failed to inhibit the redness in this model.

Lipoxygenase inhibitors may have been ineffective in the AA model because they are ineffective in the treatment of ocular surface inflammation. Another possibility, however, is that the cyclooxygenase pathway predominates in noninflamed conjunctiva, and only after insult and polymorphonucleocyte infiltration is the lipoxygenase pathway activated. Thus, testing lipoxygenase activity and inhibition in normal volunteers would not reflect their activity in clinical ocular inflammation. On the other hand, we can conclude that cyclooxygenase inhibitors are effective and can be screened in this model.

Effects of Exogenously Applied Lipid Mediators on the Conjunctiva

Lipid mediators are synthesized and released during conjunctival inflammation. They are involved in the early and late phases of the inflammatory response. Normally the conjunctiva contains numerous cell types which are capable of producing these mediators, such as mast cells, lymphocytes, leukocytes and others.

Effects of PGD_2

PGD_2 is the main prostaglandin secreted by the activated mast cell (8 ng/10^6 cells). Activation of the mast cell occurs after the binding of an allergen to two IGE surface receptors, and release of AA by the action of phospholipase A_2. PGD_2 is an important mediator in both early and late phases of immediate hypersensitivity reactions of the conjunctiva. PGD_2 levels have been found to be elevated in serum of patients with systemic mastocytosis and in tears of those with vernal conjunctivitis.

Abelson et al. [13] showed that the topical application of PGD_2 (1.0 mg/ml) to guinea pig and human eyes produces conjunctival injection, mucous discharge and eosinophil chemotaxis (fig. 6). In contrast to histamine, PGD_2 did not produce any itching in humans and the conjunctival injection involved deep episcleral conjunctival blood vessels instead of superficial ones.

Effects of LTB_4

LTB_4 is mainly recognized for its ability to stimulate leukocyte chemotaxis in vitro and chemoattraction of leukocytes and their adherence to vascular endothelium in vitro. Its application to rabbit and cat eyes induced moderate injection and marked polymorphonuclear chemotaxis in the conjunctival tissue which peaked at 3 h and enduring for up to 12 h [14] (fig. 7). Similarly, LTB_4 produced polymorphonuclear infiltration, without clinical ocular changes, but with prolonged ocular pain lasting several days in one human subject (MBA; unpubl. data). LTB_4 levels were shown to be elevated in synovial fluids of patients with rheumatoid arthri-

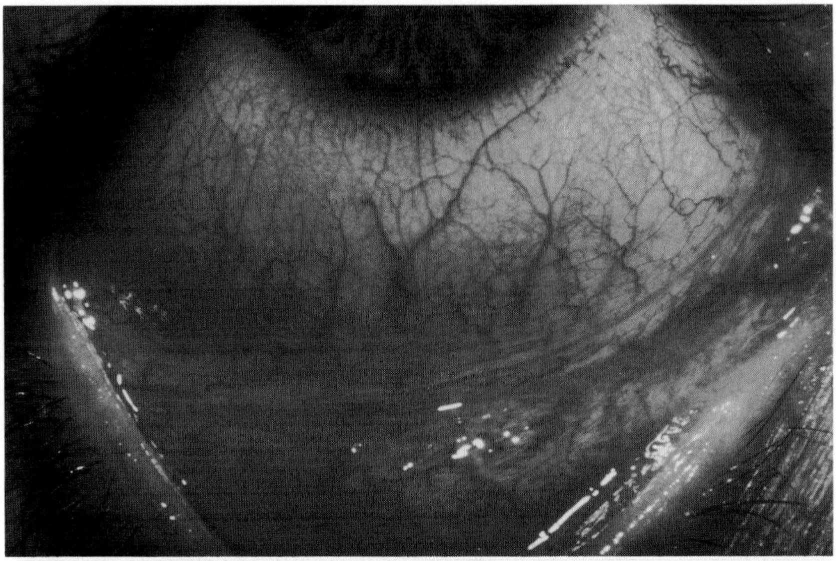

6

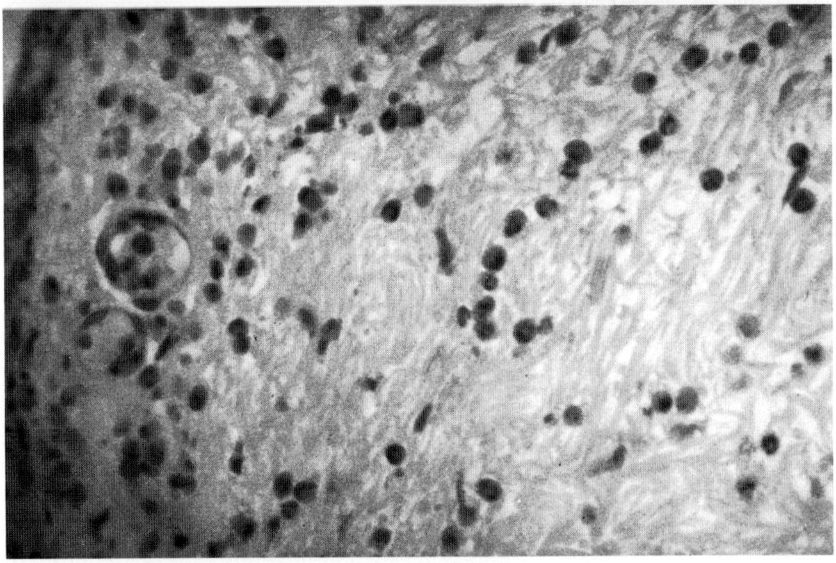

7

Fig. 6. Episcleral and limbal vessel dilation caused by one drop of PGD_2 (1 mg/ml). Mucous discharge was present but no itching.

Fig. 7. LTB_4 (100 µg/ml) effect on the rabbit conjunctiva. Extensive polymorphonuclear infiltration is apparent at 3 h.

tis and aqueous humor of those with uveitis [15]. Increase of LTB_4 levels was also present in tears of patients with vernal conjunctivitis and alkali burns [16].

Effects of LTC_4 and LTD_4

LTC_4 and LTD_4 are known for their bronchoconstrictive effects and role in asthma. They increase vascular permeability but have no chemotactic effects. They are both miotics and if applied topically to rabbit and human eyes, no injection, pain or chemotaxis were obtained [unpubl. data]. Conversely, inhalation of LTC_4 and LTD_4 by humans does cause bronchoconstriction [17, 18] and intracutaneous injection of LTC_4 and LTD_4 causes a wheal and flare that persists for 2–6 h [19, 20]. Subconjunctival injections of LTC_4 and LTD_4 were not attempted in rabbits or humans. Although the topical effects of LTC_4 and LTD_4 on the conjunctiva showed no clinical changes, tear levels of LTB_4, LTC_4, PGD_2 and PGE_2 were high in patients with vernal conjunctivitis [16] and active cicatricial pemphigoid [4].

Lipid Mediators' Release after Conjunctival Provocation by Allergens

In vivo allergen provocation experiments performed in the skin, nasal mucosa and conjunctiva, demonstrated the release of mainly LTC_4 and PGD_2 into skin chambers [21, 22], nasal washings [23] and tear fluids [24, 25]. This release is coincident with the release of other mast cell markers such as tryptase and histamine.

Conjunctival provocation is performed by instilling the lowest dose of an allergen in a subject's eye and phosphate-buffered saline (PBS) in the other eye. The allergen can be either ragweed, grass or cat dander depending on the subject's skin testing or history of previous allergy.

A positive clinical response to that particular allergen consists of moderate to severe itching and conjunctival redness (fig. 8). In those subjects in whom the clinical response is negative, the next higher dose of allergen is instilled 10 min later. The 10-min cycles are repeated until either a positive clinical response is obtained, or the highest allergen concentration is achieved.

In the skin and nasal systems, LTC_4 release is allergen dose dependent and could be detected for as long as 4 h after challenge. In the conjunctival system, Bisgaard et al. [25] detected LTC_4-like immunoreactivity in the

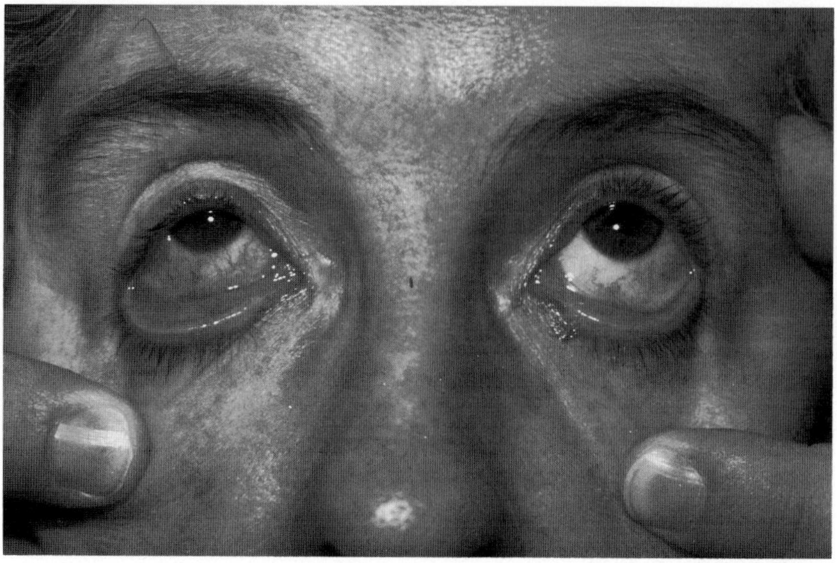

Fig. 8. A volunteer who received 19 AU/ml of ragweed allergens in the right eye and PBS in the left. Note 3+ conjunctival redness in the right eye. 3+ itching was experienced also.

tear fluid after challenge. They were able to inhibit this release by pretreatment with topical cromolyn or prednisolone. These findings underscore the importance of the role of lipid mediators in conjunctival inflammation. The failure of LTC_4 and LTD_4 to induce inflammation in the normal conjunctiva supports the theory that the lipoxygenase pathway is only activated in inflamed, and not normal, conjunctiva.

Therapeutic Implications

Clinical studies showed the efficacy of different cyclooxygenase inhibitors in the treatment of many conjunctival inflammatory disorders. Lipoxygenase inhibitor trials, however, are still in the experimental stages. Aspirin, suprofen and tolemtin proved to be beneficial in the treatment of vernal and giant papillary conjunctivitis. They have a mild therapeutic role in scleritis and episcleritis, in which the conjunctiva is secondarily involved. Further investigations of new anti-inflammatory agents are needed, especially in light of the need for a substitute for topical steroids.

References

1 Abelson MB, Udell IJ: Ocular Allergy Update. Amsterdam, Excerpta Medica, 1981.
2 Smith RE, Schanzlin DG: Rheumatoid disease; in Smolin G, Thoft RA (eds): The Cornea: Scientific Foundations and Clinical Practice. Boston, Little Brown, 1983, pp 249–263.
3 Foster CS: Ocular manifestations of the nonrheumatic acquired collagen vascular diseases; in Smolin G, Thoft RA (eds): The Cornea: Scientific Foundations and Clinical Practice. Boston, Little Brown, 1983, pp 264–284.
4 Foster CS: Cicatricial pemphigoid. Am Ophthalmol Soc 1986;84:527–663.
5 Bhattacherjee P, Kulkarni PS, Eakins, KE: Differential inflammatory effects of arachidonic acid on rabbit conjunctiva and iris: A possible role of lipoxygenase in the conjunctival response. Adv Prostaglandin Thromboxane Res 1980;8:1727–2731.
6 Bhattacherjee P, Eakins KE: Inhibition of the ocular effects of sodium arachidonate by anti-inflammatory compound. Prostaglandins 1975;9:175–182.
7 Podos SM, Becker, B, Kass MA: Prostaglandin synthesis inhibition and intraocular pressure. Invest Ophthalmol Vis Sci 1973;12:426–433.
8 Abelson MB, Butrus SI, Weston JH: Aspirin therapy in vernal conjunctivitis. Am J Ophthalmol 1983;95:502–505.
9 Buckley DC, Caldwell DR, Reaves TA Jr: Treatment of vernal conjunctivitis with suprofen, a topical non-steroidal anti-inflammatory agent. Invest Ophthalmol Vis Sci 1986;27(suppl):29.
10 Wood TS, Stewart RH, Boman RW, et al.: Suprofen treatment of contact lens-associated giant papillary conjunctivitis. Ophthalmol 1988;95:822–816.
11 Meyer E, Kraus E, Zonis S: Efficacy of antiprostaglandin therapy in vernal conjunctivis. Br J Ophthalmol 1987;71:497–499.
12 Abelson MB, Butrus SI, Kliman GH, et al: Topical arachidonic acid: A model for screening anti-inflammatory agents. J Ocul Pharmacol 1987;3:63–75.
13 Abelson MB, Madiwale NA, Weston JH: The role of prostaglandin D_2 in allergic ocular disease; in O'Connor GR, Chandler JW (eds): Advances in Immunology and Immunopathology of the Eye. New York, Masson, 1985, pp 163–166.
14 Butrus SI, Corey EJ, Weston JH, et al: The effect of leukotriene B_4 in rabbit and guinea pig eyes. Invest Ophthamol Vis Sci 1984;25(suppl):109.
15 Parker JA, Goetzl EJ, Friedlaender MH: Leukotrienes in the aqueous humor of patients with uveitis. Arch Ophthalmol 1986;104:722–724.
16 Abelson MD: Lipoxygenase products in ocular inflammation. Invest Ophthalmol Vis Sci 1984;25(suppl):42.
17 Weiss JW, Frazen JM, Coles N, et al: Bronchoconstrictor effects of leukotriene C in humans. Science (Wash) 1982;26:196–198.

18 Holroyd MD, Altonyan REC, Cole M, et al: Bronchoconstriction produced in man by leukotrienes C and D. Lancet 1981;ii:17–18.
19 Soter NA, Lewis RA, Corey EJ, et al: Local effects of synthetic leukotrienes (LTC_4, LTD_4, LTE_4 and LTB_4) in human skin. J Invest Dermatol 1983;80:115–119.
20 Bisgaard HJ, Kristensen J, Sondengaard J: The effect of leukotriene C_4 and D_4 on cutaneous blood flow in humans. Prostaglandins 1982;23:797–801.
21 Talbot SF, Atkins PC, Goetzel EJ, et al: Accumulation of leukotriene C_4 and histamine in human allergic skin reactions. J Clin Invest 1985;76:650–656.
22 Pinekowski MM, Adkinson NF Jr, Plant M, et al: Prostaglandin D_2 and histamine during the immediate and the late phase components of allergic cutaneous responses. J Allergy Clin Immunol 1988;82:95–100.
23 Creticos PS, Peters SP, Adkinson NF Jr, et al: Peptide leukotriene release after antigen challenge in patients sensitive to ragweed. N Engl J Med 1984;310:1626–2630.
24 Bisgaard H, Ford-Hutchinson AW, Charleson S, et al: Detection of leukotriene C_4-like immunoreactivity in tear fluid from subjects challenged in the specific allergen. Prostaglandins 1984;27:369–374.
25 Bisgaard H, Ford-Hutchinson AW, Charleson S: Production of peptido-lipid leukotrienes in human tear fluid following antigen challenge. Prostaglandins 1984;28:620–622.

Salim I. Butrus, MD, George Washington University Medical Center,
Washington, DC 20037 (USA)

Bazan NG (ed): Lipid Mediators in Eye Inflammation.
New Trends Lipid Mediators Res. Basel, Karger, 1990, vol 5, pp 41–55

Leukotrienes and Allergic Conjunctivitis

Chi-Chung Chan [1]

Department of Pharmacology, Merck Frosst Center for Therapeutic Research, Pointe-Claire-Dorval, Que., Canada

Leukotrienes (LTs) are a group of potent mediators derived from the 5-lipoxygenase pathway of arachidonic acid metabolism. LTB_4 is derived from arachidonic acid through the actions of two enzymatic steps, initially via 5-lipoxygenase to form the unstable intermediate LTA_4, followed by subsequent conversion to LTB_4 by LTA_4 hydrolase. LTB_4 has potent chemotactic and chemokinetic activities for polymorphonuclear leukocytes through activation of high affinity and structurally specific receptor sites. LTC_4, LTD_4 and LTE_4 are peptidolipid conjugates derived from LTA_4 through conjugation with glutathione. These peptidolipid conjugates have potent bronchoconstrictor actions in humans and are now known to be the constituents of slow-reacting substances of anaphylaxis (SRS-A). Recent studies have suggested that LTs, in particular peptidolipid LTs, also have potent actions in the eye and may be involved in ocular inflammatory responses and immediate hypersensitivity reactions in the conjunctiva. The aim of this review is to summarize the evidence for the involvement of LTs in allergic conjunctivitis and explore the room for the development of either LT antagonists or inhibitors as potential treatments for allergic conjunctivitis.

Actions of Leukotrienes in the Eye

The eye has been a neglected organ in LT research until recently. Early work had centered on the possible relationship between LTs and changes in intraocular pressure. It is now known that LTs have little if any effect in

[1] I thank Dr. Anthony Ford-Hutchinson, Philip Tagari and Aidan Foster for helpful discussions and suggestions in the preparation of the manuscript.

this respect [1–4]. More recent studies have concentrated on the role of LTs in immediate hypersensitivity reactions in the conjunctiva. LTs indeed have potent effects in the conjunctiva in recruiting inflammatory cells and enhancing vascular permeability. A considerable amount of work has been carried out by Woodward and co-workers. Table 1 summarizes the pharmacological actions of LTB_4, and peptidolipid LTs in different species.

When applied either topically or injected directly into the anterior chamber of the eye of the guinea pig, cat and rabbit, LTB_4 consistently induced leukocyte/neutrophil accumulation in the conjunctiva or the anterior chamber. In the guinea pig, LTB_4 at higher doses also recruited eosinophils in the conjunctiva [5, 6], in agreement with other observations in vitro and in vivo [7–10]. Neither LTB_4 nor the peptidolipid LTs have any significant effect on intraocular pressure in a number of species [1, 3]. In contrast to the peptidolipid LTs, LTB_4 has no effect on vascular permeability changes in the conjunctiva in the guinea pig [11].

LTD_4, LTC_4 and LTE_4 have potent vascular permeability-enhancing actions in the conjunctiva of the guinea pig and the hamster, as measured by extravasation of radioactive macromolecules such as human or bovine serum albumin. The rank order of potencies in the guinea pig was $LTE_4 \geq LTD_4 > LTC_4$ [12] whereas in the hamster it was $LTD_4 > LTC_4 >> LTE_4$ [11]. The actions of LTD_4 were stereospecific, the activity of the natural 5S,6R-LTD_4 stereoisomer being 100–1,000 times than that of the 5R,6S-LTD_4 stereoisomer [11, 12]. Interestingly, LTD_4 induced increases in vascular permeability in the guinea pig regardless of whether it was administered topically [12, 13] or intravenously [14]. In both cases, the responses to LTD_4 were not affected by pretreatment with indomethacin, suggesting that cyclooxygenase metabolites were not involved in the process. This is in marked contrast to the observations in the guinea pig lung where it has been shown that bronchoconstriction produced by intravenously administered, but not aerosolized, LTD_4 could be significantly inhibited by either indomethacin or a thromboxane synthetase inhibitor [15]. Thus it is quite clear that in the guinea pigs intravenously administered LTD_4 can induce the release of prostaglandins and thromboxanes. These observations also provide circumstantial evidence that neither prostaglandins nor thromboxanes have any significant permeability-enhancing effects in the guinea pig conjunctiva. In fact, it has been shown that direct application of prostaglandins (E_1, E_2 and $F_{2\alpha}$) in the guinea pig conjunctiva produces no observable inflammatory responses [16].

Table 1. Effects of LTs in the eye

Species	Leuko-trienes	Mode of administration	Dose, nmol	Effect	Ref.
Guinea pig	LTB_4	injected into the anterior chamber	1.6	leukocyte accumulation in aqueous humor	3
		topical	0.34–3.4	neutrophil/eosinophil infiltration in conjunctiva	5, 6
	LTC_4	topical	0.16–1.6	increase conjunctival vascular permeability	12
	LTD_4	topical	0.02–2	eosinophil infiltration in conjunctiva	5, 6
		topical	0.002–20	increase conjunctival vascular permeability	12, 13
		intravenous	0.3–3 nmol/kg	increase conjunctival vascular permeability	14
	LTE_4	topical	0.02–2.3	eosinophil infiltration in conjunctiva	5
		topical	0.02–23	increase conjunctival vascular permeability	12
Hamster	LTC_4	topical	0.08–8	increase conjunctival vascular permeability	11
	LTD_4	topical	0.01–10	increase conjunctival vascular permeability	11
	LTE_4	topical	0.11–11	increase conjunctival vascular permeability	11
Cat	LTB_4	injected into the anterior chamber	4.7	leukocyte accumulation in aqueous humor	3
	LTC_4 LTD_4	injected into the anterior chamber	0.16–1.6 0.2–2	miosis	3, 4
Rabbit	LTB_4	injected into the anterior chamber	0.6–1.2	leukocyte accumulation in aqueous humor	1–3

LTs also have chemotactic activities for eosinophils, a cell type which has been suggested to play a significant role in immediate hypersensitivity reactions [17–19]. Both LTD_4 and LTE_4 induced eosinophil infiltration into the guinea pig conjunctiva at doses considerably lower than that required for LTB_4 [5, 6]. The optimum infiltration was observed at 4–6 h after topical administration of the LTs, which is comparable to the time course of antigen-induced eosinophil migration in the guinea pig lung [20]. The eosinophil chemotactic effects do not appear to be related to the vasopermeability-enhancing actions of the peptidolipid LTs since bradykinin, which has potent vasopermeability-enhancing actions, has no effect at all on eosinophil infiltration in the conjunctiva. Preliminary results from our laboratory also suggested that peptidolipid LTs have eosinophil chemotactic activity in the guinea pig lung [Foster and Chan, results to be published]. Spada et al. [6] have demonstrated in an elegant study that LTB_4 synergizes with LTD_4 in the recruitment of eosinophils and neutrophils in the guinea pig conjunctiva, via mechanism(s) unrelated to their potent vasopermeability actions. This interaction between LTB_4 and LTD_4 would be especially important in the early stage of allergic responses in that it may optimize the recruitment of eosinophils and neutrophils in the conjunctiva.

For obvious ethical reasons, studies on the action of LTs in human eye have been limited. The only study, reported in abstract form, showed that instillation of LTC_4 into the eyes of two subjects produced no observable clinical effect [21]. However, these results are far from conclusive.

Generation of Lipoxygenase Products in Ocular Tissues

The ocular tissues of many species, from guinea pig to humans, have the capacity to generate a variety of lipoxygenase products (table 2). As has been observed for the lung [22], the skin [23] and the gastrointestinal tract [24], ocular tissues are able to convert arachidonic acid to different LTs and other lipoxygenase products, suggesting the presence of the 5- and 12-lipoxygenase. In addition, challenge with the calcium ionophore A23187 in either the conjunctiva of the guinea pig or the rabbit, the anterior uvea of the rabbit or the bovine lens resulted in increased production of LTB_4 and peptidolipid LTs [25–29]. Currently the guinea pig conjunctiva and anterior uvea are the only ocular tissues from animals which have

Table 2. Generation of lipoxygenase metabolites in ocular tissues

Species	Tissue	Lipoxygenase metabolites generated	Stimulation or substrate used	Ref.
Guinea pig	conjunctiva	LTB_4	antigen or A23187 challenge	25
	conjunctiva anterior uvea	SRS-A	antigen or A23187 challenge	30
	conjunctiva iris	12-HETE	radiolabeled AA	39
Rat	conjunctiva	5-HETE 12-HETE	radiolabeled AA	39
Rabbit	anterior uvea	5-HETE LTB_4	radiolabeled AA	26, 27
	anterior uvea conjunctiva	SRS-A	A23187 challenge	26, 27
	iris aqueous humor retina	LTB_4	injection of endotoxin into vitreous cavity or A23187 challenge	40, 45, 46
	anterior uvea conjunctiva cornea	12-HETE	radiolabeled AA	26, 37–40, 42, 47
Bovine	lens	LTB_4	A23187 challenge or radiolabeled AA	28, 29
Monkey	anterior uvea conjunctiva eyelids	LTB_4 5-HETE 12-HETE	radiolabeled AA	38, 39
Human	lens	LTB_4	radiolabeled AA	29
	anterior uvea	LTB_4 5-HETE 12-HETE	radiolabeled AA	48
	aqueous humor	LTB_4 LTC_4, LTD4, LTE_4	in patients with uveitis	31
	tear fluid	LTC_4, LTD_4, LTE_4	allergen challenge	32, 33

been clearly shown to respond to antigen challenge with elevated production of either LTB_4 [25] or peptidolipid LTs [30].

In humans, significant amounts of LTB_4 and $LTC_4/LTD_4/LTE_4$ were present in the aqueous humor of control subjects as well as patients with uveitis, the amount of the peptidolipid LTs being 2–3 times that of LTB_4 [31]. The most relevant finding may be that elevated amounts of LTC_4 and LTD_4 were recovered in tear fluid in allergic individuals upon specific allergen challenge [32, 33], this coinciding with the development of clinical symptoms such as redness, swelling and itching. The sources of the LTC_4 and LTD_4 are likely to be the mast cells and eosinophils as human mast cells [34] and eosinophils [35, 36] have been shown to be capable of producing LTs. These observations suggest that peptidolipid LTs may be involved in the human ocular allergic responses.

12-Lipoxygenase products are also generated in the conjunctiva and anterior uvea in a number of species. In fact, the percentage conversion from ^{14}C-arachidonic acid to 12-HETE is consistently greater than that to 5-lipoxygenase products in the rabbit conjunctiva and the inflamed rabbit cornea [26, 37–40]. The stereochemistry of the 12-HETE produced has not been determined. 12(S)-HETE can be produced by 12-lipoxygenase whereas 12(R)-HETE can be produced by a cytochrome P-450-dependent process. These two products have important differences in biological activities. 12(R)-HETE has similarities to LTB_4 in chemical structure and, like LTB_4, has potent neutrophil chemotactic activity [41]. It has been reported that 12(R)-HETE and to a much less extent 12(S)-HETE displace bound LTB_4 from rat and human neutrophil membrane, suggesting that they share the same receptor sites. Thus it is possible that 12-HETE may also play a role in inflammatory reponses in the conjunctiva. In other tissue such as the rabbit cornea, it has been suggested that 12-HETE may be involved in the wound-healing process [37, 42]. In this regard, it is of interest to note that 12-HETE, particularly 12(R)-HETE, has been shown to have proliferative activities in the epidermis [43, 44].

Animal Model of Allergic Conjunctivitis

There is no ideal animal model for human allergic conjunctivitis per se. However, one can measure in animal models specific parameters such as permeability responses and infiltration of inflammatory cells which are characteristic features in the human disease. The guinea pig is a convenient and perhaps a useful model for the study of the role of LTs in allergic conjunctivitis for a number of reasons. As mentioned above, LTs produce significant and easily measurable permeability responses and leukocyte infiltration in the guinea pig conjunctiva. Indeed, it has been demonstrated that LTB_4 is released in the conjunctiva following antigen challenge. In addition, the guinea pig model has been reasonably well characterized [16, 49–56]. Either local injection or topical instillation of specific antigen to the conjunctiva of sensitized animals has produced responses such as irritation, edema, hyperemia, production of watery discharge, degranulation of mast cells, infiltration of neutrophils and eosinophils; features morphologically and histologically similar to that observed in the human disease. In contrast, antigen challenge produced no observable responses in the conjunctiva of either the rat (CFHB strain), the rabbit or a number strains of mouse [16]. Although antigen-induced responses have been demonstrated in the Sprague-Dawley rat [57, 58], the rat neutrophil, unlike the guinea pig neutrophil, does not show a chemotactic response to LTB_4 [59]. Hence the rat may not be an appropriate model for the study of inflammatory reactions involving LTB_4.

The conjunctiva of the gunea pig, but not that of either the mouse, the rat or the rabbit, has been shown to respond with strong reactions to topical challenge with histamine [16], also an important mediator in human allergic conjunctivitis. Thus H_1 antagonists which are effective in the treatment of allergic conjunctivitis are also consistently effective in reducing the inflammatory responses in the guinea pig conjunctiva [16, 25, 52]. Disodium cromoglycate has also been tested in this model, with somewhat mixed results varying from palliative [16, 53], ineffective [52] and aggravation [53]. This may relate to the fact that disodium cromoglycate fails to work in other models of immediate hypersensitivity in the guinea pig and it has been suggested that there is species difference [60, 61].

LTD_4 Receptor Antagonists and Leukotriene Synthesis Inhibitors

Since the initial work on the chemical and biochemical characterization of LTs has suggested that these mediators may be of pathological significance, considerable effort has been expended in finding specific chemical molecules to modify the actions of LTs as a potential treatment for diseases such as asthma, inflammatory bowel diseases, psoriasis and allergic conjunctivitis [see 62 for recent reviews and studies]. Prototypic

LTD_4 receptor antagonists and 5-lipoxygenase inhibitors, which have potency in the millimolar range, have been available for several years [63–68]. However, the preliminary clinical efficacy studies with these compounds have been disappointing [69, 70], perhaps reflecting their low potency. Recently, more potent compounds, which either antagonize the actions of LTD_4 or inhibit 5-lipoxygenase at the nanomolar level (classified here as second-generation antagonists and inhibitors) have become available [71–75]. The effects of the prototypic and the second-generation compounds have been examined in the guinea pig model of allergic conjunctivitis, with excellent correlation between their in vitro potency and in vivo activity.

The classical SRS-A antagonist FPL 55712 and the prototypic LTD_4 antagonists L-648,051, L-649,923 and SKF 102922 have been demonstrated to have specific inhibitory effect against LTD_4-induced vascular permeability increases in the guinea pig conjunctiva [12, 14, 74]. These compounds also effectively inhibit conjunctival responses to a single antigen challenge when the histamine component is blocked by histamine antagonist. The responses were inhibited by a total of 80%, suggesting that LTs and histamine together account for a major part of the antigen-induced permeability increases. The LTD_4 antagonists had little or no effect on antigen challenge when the histamine component was left intact. On the other hand, if the guinea pigs were given a second antigen challenge 24 h after the first, administration of the antagonists produced a significant inhibition [14]. Histological studies reveal that a large number of eosinophils were present in the conjunctiva 24 h after the first antigen challenge. As eosinophils can produce LTs, this may explain why LTD_4 antagonists were more effective against a double than a single antigen challenge.

As expected, a second-generation LTD_4 antagonist MK-571 (also known as L-660,711) [71] was 100–1,000 times more potent in the guinea pig conjunctiva. It has been shown that the vascular permeability response to a single antigen challenge was significantly inhibited by MK-571 [76] and another LTD_4 antagonist, SKF 104353 [74]. These results suggested that peptidolipid LTs are involved in the responses to a single antigen challenge in the guinea pig conjunctiva and that the failure of the prototypic antagonists to block such responses was due to their low potency. Interestingly, treatment with either MK-571 or a combined histamine H_1 and H_2 antagonist both inhibit about 75% of the conjunctival responses [76]. One of the explanations may be that there is potentiating interaction between the LT and histamine components following antigen challenge. Thus, pharmacological antagonism of either LTD_4 or histamine alone produce maximal effects on the conjunctival responses.

5-Lipoxygenase inhibitors are more efficacious than the LTD_4 antagonists in the guinea pig conjunctiva [25], suggesting that other mechanisms mediated through either LTB_4 or LTC_4 receptors are involved and that LTB_4 may interact with the peptidolipid LTs. Indeed, it has been demonstrated that LTB_4 potentiates the action of LTD_4 in eosinophil and neutrophil migration in the guinea pig conjunctiva [6]. A second-generation inhibitor MK-886 (also known as L-663,536) [73] has an ED_{50} value of 0.05 mg/kg (i.v.) against the permeability responses to a single antigen challenge [Chan et al., unpubl. results], again showing improved potency in the conjunctiva compared to that of the prototypic inhibitors.

The anti-inflammatory effects of a number of compounds having anti-LT actions have been shown in other models of ocular inflammation. The list of compounds includes the anti-SRS-A compound KC-404 [30], and the nonspecific inhibitors REV 5901 [77, 78], Bay 08276 [78], BW 755C [79], TEI-1338 [46], TEI-1345 [46], nafazatrom [80], phenidone [81], nordihydroguaiaretic acid [80, 81] and 5,8,11,14-eicosatetraenoic acid [82]. All of these compounds reduce experimental inflammation in the eye to a significant extent. It should be noted that some of these compounds have complex mechanisms of action, hence their anti-inflammatory effects may not necessarily be related to their anti-LT actions.

Potential Treatment for Allergic Conjunctivitis

Currently, allergic conjunctivitis in man is treated with either H_1 antagonists, disodium cromoglycate or topical corticosteroids. H_1 antagonists are effective in mild forms of allergic conjuctivitis but are less effective in chronic situations which may respond to either disodium cromoglycate or corticosteroids. Disodium cromoglycate has been described as an inhibitor of mediator release acting through 'mast cell-stabilizing effects' [60, 61, 83]. While it is unknown whether such a mechanism is related to the therapeutic effect of disodium cromoglycate in the eye, it is quite clear that other mechanisms are involved in the therapeutic effect of disodium cromoglycate in asthma. Many compounds which are more potent than disodium cromoglycate as mast cell-stabilizing agents have been developed [83] but none of them are useful in the treatment of asthma.

Although corticosteroids are very effective in the treatment of allergic conjunctivitis, they are not widely used because of potential side effects. Some investigators indeed suggest that topical steroids are rarely needed for the treatment of allergic conjunctivitis [84]. Thus, there is a need for the development of novel nonsteroidal anti-inflammatory agents which have a clearly defined mechanism of action for the treatment of allergic conjunctivitis. LTD_4 receptor antagonists and LT biosynthesis inhibitors are strong candidates in view of the evidence for the involvement of LTs in this disease and the effectiveness of these compounds in experimental models of allergic conjunctivitis and ocular inflammation. Since LTs and histamine are two major mediators involved and there is evidence that they interact with each other, LTD_4 antagonists and LT biosynthesis inhibitors may be of potential therapeutic use when administered either alone or in combination with H_1 antagonists.

At the present time little information is available with regard to the clinical efficacy of either LT antagonists or inhibitors. A single study [85] has been carried out with AA-673, a compound with weak LTD_4 antagonistic action [86, 87], in which it was demonstrated that a 0.25% ophthalmic solution of this compound was as effective as disodium cromoglycate in the treatment of allergic and vernal conjunctivitis (it has also been reported that this compound was effective in the treatment of allergic rhinitis [88]). However, AA-673 also has other actions, such as an inhibitory action on histamine release [86, 87]. Thus a clear link between antagonism of LKT action and the therapeutic effect of AA-673 cannot be established from this study.

Summary and Conclusions

There is now strong evidence in support of a role of LTs in allergic conjunctivitis. Antigen challenge in either human or animal eyes results in the release of LTs which have potent vascular permeability-enhancing and leukocyte chemotactic effects in the conjunctiva. Specific LTD_4 receptor antagonist and LT biosynthesis inhibitors are as effective as histamine H_1 and H_2 antagonists in reducing the immediate hypersensitivity reactions in the guinea pig conjunctiva. The effectiveness of the LT antagonists and inhibitors correlates very well with their anti-LT activities. It will thus be of considerable interest to ascertain that these compounds are also effective in the treatment of allergic conjunctivitis in man.

References

1 Bhattacherjee P, Hammond B, Salmon J, et al: Chemotactic response to some arachidonic acid lipoxygenase products in the rabbit eye. Eur J Pharmacol 1981;73: 21–28.
2 Bhattacherjee P, Hammond B, Salmon J, et al: Effect of lipoxygenase products on leukocyte accumulation in the rabbit eye. Adv Prostaglandin Thromboxane Leukotriene Res 1982;9:325–330.
3 Stjernschantz J, Sherk T, Borgeat P, et al: Intraocular effects of lipoxygenase pathway products in arachidonic acid metabolism. Acta Ophthalmol 1984;62:104–111.
4 Stjernschantz J, Sherk T, Sears M: Ocular responses to leukotriene C_4 and D_4 in the cat. Prostaglandins 1984;27:5–15.
5 Spada CS, Woodward DF, Hawley SB, et al: Leukotrienes cause eosinophil emigration into conjunctival tissue. Prostaglandins 1986;31:795–809.
6 Spada CS, Woodward DF, Hawley SB, et al: Synergistic effects of LTB_4 and LTD_4 on leukocyte emigration into the guinea pig conjunctiva. Am J Pathol 1988;130: 354–368.
7 Czarnetzki BM, Rosenbach T: Chemotaxis of human neutrophils and eosinophils towards leukotriene B_4 and its 20-ω-oxidation products in vitro. Prostaglandins 1986;31:851–858.
8 Palmblad J, Uden AM, Friberg I, et al: Effects of novel lipoxygenase products on neutrophil and eosinophil functions in vitro. Adv Prostaglandin Thromboxane Leukotriene Res 1983;12:25–311.
9 Movat HZ, Rettl C, Burrowes CE, et al: The in vivo effect of leukotriene B_4 on polymorphonuclear leukocytes and the microcirculation. Comparison with activated complement (C5a-des-Arg) and enhancement by prostaglandin E_2. Am J Pathol 1984;115:233–244.
10 Hakansson L, Westerlund D, Venge P: New method for measurement of eosinophil migration. J Leukocyte Biol 1987;42:689–696.
11 Woodward DF, Ledgard SE: Comparison of leukotrienes as conjunctival microvascular permeability factors. Ophthalmic Res 1985;17:318–320.
12 Gary RK, Woodward DF, Nieves AL, et al: Characterization of the conjunctival vasopermeability response to leukotrienes and their involvement in immediate hypersensitivity. Invest Ophthalmol Vis Sci 1988;29:119–126.
13 Woodward DF, Ledgard SE: Effect of LTD_4 on conjunctival vasopermeability and blood-aqueous barrier integrity. Invest Opthalmol Vis Sci 1985;26:481–485.
14 Garceau D, Ford-Hutchinson AW: The role of leukotriene D_4 as a mediator of allergic conjunctivitis in the guinea-pig. Eur J Pharmacol 1987;134:285–292.
15 Hamel R, Masson P, Ford-Hutchinson AW, et al: Differing mechanisms for leukotriene D_4-induced bronchoconstriction in guinea pigs following intravenous and aerosol administration. Prostaglandins 1982;24:419–432.
16 Tuffin DP, Dayhimi I, Feinberg JG: Immunopharmacology of the guinea pig conjunctiva. Ann Immunol Hung 1986;26:601–628.
17 Butterworth AE, David JR: Eosinophil function. New Engl J Med 1981;304:154–156.
18 De Monchy JGR; Kauffman HF, Venge P, et al: Bronchoalveolar eosinophilia during allergen-induced late asthmatic reactions. Am Rev Respir Dis 1985;131:373–376.

19 Kay AB: Eosinophils as effector cells in immunity and hypersensitivity disorders. Clin Exp Immunol 1985;62:1–12.
20 Dunn CJ, Elliott GA, Oostveen JA, et al: Development of a prolonged eosinophil-rich inflammatory leukocyte infiltration in the guinea-pig asthmatic response to ovalbumin inhalation. Am Rev Respir Dis 1988;137:541–547.
21 Weston JH, Abelson MB: Leukotriene C_4 in rabbit and human eyes. Invest Ophthalmol Vis Sci 1985;26:191.
22 Dahlen S-E, Hansson G, Hedqvist P, et al: Allergen challenge of lung tissue from asthmatics elicits bronchial contraction that correlates with the release of leukotrienes C_4, D_4, and E_4. Proc Natl Acad Sci USA 1983;80:1712–1716.
23 Ziboh VA, Casebolt TL, Marcelo CL, et al: Biosynthesis of lipoxygenase products by enzyme preparations from normal and psoriatic skin. J Invest Dermatol 1984;83:426–430.
24 Dreyling KW, Hoppe U, Peskar BA, et al: Leukotriene synthesis by human gastrointestinal tissues. Biochim Biophys Acta 1986;878:184–193.
25 Garceau D, Ford-Hutchinson AW, Charleson S: 5-Lipoxygenase inhibitors and allergic conjunctivitis reactions in the guinea-pig. Eur J Pharmacol 1987;143:1–7.
26 Kulkarni PS, Srinivasan BD: Synthesis of slow-reacting substance-like activity in rabbit conjunctiva and anterior uvea. Invest Ophthalmol Vis Sci 1983;24:1079–1085.
27 Kin K: Release of slow-reacting substance from rabbit ocular tissues. Jap J Ophthalmol 1984;28:263–273.
28 Fujiwara H, Nakata K: Biosynthesis of leukotriene B_4 in bovine lens. Jap J Ophthalmol 1988;32:44–50.
29 Fujiwara H, Nakata K: The biosynthesis of leukotriene B_4 in human cataractous lens and in bovine lens. Acta Soc Ophthalmol Jap 1985;89:1182–1187.
30 Sunada A, Kin K, Narumi C, et al: The role of leukotrienes in ocular inflammation report 2. The effect of anti-SRS-A substance (KC-404) on experimental uveitis in guinea pigs. Acta Soc Ophthalmol Jap 1984;88:1341–1349.
31 Parker JA, Goetzl EJ, Friedlaender MH: Leukotrienes in the aqueous humor of patients with uveitis. Arch Ophthalmol 1986;104:722–724.
32 Bisgaard H, Ford-Hutchinson AW, Charleson S, et al: Production of peptido-lipid leukotrienes in human tear fluid following antigen challenge. Prostaglandins 1984;28:620–622.
33 Bisgaard H, Ford-Hutchinson AW, Charleson S, et al: Production of leukotrienes in human skin and conjunctival mucosa after specific allergen challenge. Allergy 1985;40:417–423.
34 MacGlashan DW, Schleimer RP, Peters SP, et al: Generation of leukotrienes by purified human lung mast cells. J Clin Invest 1982;70:747–751.
35 Weller PF: Functional and biochemical properties of selected protein constituents of human eosinophils; in Yoshida T, Torisu M (eds): Immunobiology of the Eosinophil. New York, Elsevier, 1983, pp 145–161.
36 Jorg A, Henderson WR, Murphy RC, et al: Leukotriene generation by eosinophils. J Exp Med 1982;155:390–402.
37 Bazan HEP, Birkle DL, Beuerman RW, et al: Inflammation-induced stimulation of the synthesis of prostaglandins and lipoxygenase-reaction products in rabbit cornea. Curr Eye Res 1985;4:175–179.

38 Kulkarni PS, Kaufman PL, Srinivasan BD: Cyclo-oxygenase and lipoxygenase pathways in cynomolgus and rhesus monkey conjunctiva, anterior uvea and eyelids. Curr Eye Res 1987;6:801–808.
39 Williams RN, Bhattacherjee P, Eakins KE: Biosynthesis of lipoxygenase products by ocular tissues. Exp Eye Res 1983;36:397–402.
40 Preud'homme Y, Demolle D, Boeynaems JM: Metabolism of arachidonic acid in rabbit iris and retina. Invest Ophthalmol Vis Sci 1985;26:1336–1342.
41 Evans JF, Leblanc Y, Fitzsimmons B, et al: Activation of leukocyte movement and displacement of [^{3}H]leukotriene B_4 from leukocyte membrane preparations by (12R)- and (12S)-hydroxyeicosatetraenoic acid. Biochim Biophys Acta 1987;917: 406–410.
42 Birkle DL, Sanitato JJ, Kaufman HE, et al: Arachidonic acid metabolism to eicosanoids in herpes virus-infected rabbit cornea. Invest Ophthalmol Vis Sci 1986;27: 1443–1446.
43 Chan C-C, Duhamel L, Ford-Hutchinson AW: Leukotriene B_4 and 12-hydroxyeicosatetraenoic acid stimulate epidermal proliferation in vivo in the guinea pig. J Invest Dermatol 1985;85:333-334.
44 Ruzicka T, Burg G: Effects of chronic intracutaneous administration of arachidonic acid and its metabolites – Induction of leukocytoclastic vasculitis by leukotriene B_4 and 12-hydroxyeicosatetraenoic acid and its prevention by prostaglandin E_2. J Invest Dermatol 1987;88:120–123.
45 Fujiwara H, Takigawa Y, Katayama T, et al: The significance of leukotriene B_4 in experimental and clinical uveitis. Acta Soc Ophthalmol Jap 1988;92:209–214.
46 Fujiwara H, Katayama T, Nakata K, et al: Inhibitors of leukotriene B_4 (LTB_4) and their effects on ocular tissues. Atarashii Ganka (J Eye) 1988;5:143–147.
47 Williams RN, Delamere NA, Peterson CA: Generation of lipoxygenase products in the avascular tissues of the eye. Exp Eye Res 1985;41:733–738.
48 Kulkarni PS, Rodriguez AV, Srinivasan BD: Human anterior uvea synthesizes lipoxygenase products from arachidonic acid. Invest Ophthalmol Vis Sci 1984;25: 221–223.
49 Feinberg JG, Chopra SL: An atopic model in the guinea pig. Excerpta Med 1967; 144:11–13.
50 Dwyer RStC, Darougar S: Models of immediate and delayed hypersensitivity in the guinea pig conjunctiva. Trans Ophthalmol Soc UK 1971;91:451–461.
51 Dwyer RStC, Turk JL, Darougar S: Immediate hypersensitivity in the guinea pig conjunctiva. I. Characterisation of the IgE and IgG_1 antibodies involved. Int Arch Allergy Appl Immunol 1974;46:910–924.
52 Dwyer RStC, Darougar S, Jones BR: Immediate hypersensitivity in the guinea pig conjunctiva. II. Effect of treatment with antihistamines, steroids and disodium cromoglycate. Mod Probl Ophthalmol 1976;16:186–195.
53 Taylor WA, Roitt IM: Effect of disodium cromoglycate on various types of anaphylactic reaction in the guinea pig. Int Arch Allergy Appl Immunol 1973;45:795–807.
54 Schwartz AE, Stock EI: Immediate hypersensitivity in the guinea pig conjunctiva. Clin Res 1983;31:817A.
55 Khatami M, Donnelly JJ, Rockey JH: Induction and down-regulation of conjunctival type-I hypersensitivity reactions in guinea pigs sensitized topically with fluoresceinyl ovalbumin. Ophthalmic Res 1985;17:139–147.

56 Khatami M, Donelly JJ, John T, et al: Vernal conjunctivitis. Model studies in guinea pigs immunized topically with fluoresceinyl ovalbumin. Arch Ophthalmol 1984; 102:1683–1688.
57 Allansmith MR, Bloch KJ, Baird RS: Ocular anaphylaxis: induction by local injection of antigen. Immunology 1981;44:623–627.
58 Allansmith MR, Baird RS, Greiner JV, et al: Late-phase reactions in ocular anaphylaxis in the rat. J Allergy Clin Immunol 1984;73:49–55.
59 Kopp DE, Esser B, Tashoff T, et al: In vivo and in vitro assessment of the role of leukotriene B_4 as a mediator of rat cutaneous late-phase reactions. J Allergy Clin Immunol 1986;77:302–308.
60 Cox JSG: Disodium cromoglycate (FPL 670, Intal): a specific inhibitor of reaginic antibody-antigen mechanisms. 1967;216:1328–1329.
61 Sheard P, Blair AMJN: Disodium cromoglycate: activity in three in vitro models of the immediate hypersensitivity reaction in lung. Int Arch Allergy Appl Immunol 1970;38:217–224.
62 Levi R, Krell RD (eds): Biology of the Leukotrienes. New York, New York Academy of Sciences, 1988.
63 Guindon Y, Girard Y, Maycock A, et al: L-651,392, a novel, potent and selective 5-lipoxygenase inhibitor. Adv Prostaglandin Thromboxane Leukotriene Res 1987; 17:554–557.
64 Chan C-C, Dubois L, Young V: Effects of two novel inhibitors of 5-lipoxygenase, L-651,392 and L-651,896, in a guinea-pig model of epidermal hyperproliferation. Eur J Pharmacol 1987;139:11–18.
65 Jones TR, Guindon Y, Champion E, et al: L-648,051: An aerosol active leukotriene D4 receptor antagonist. Adv Prostaglandin Thromboxane Leukotriene Res 1987;17: 1012–1017.
66 Jones TR, Young R, Champion E, et al: L-649,923, sodium (βS*,γR*)-4-[3-(4-acetyl-3-hydroxy-2-propylphenoxy)-propylthiol-γ-hydroxy-β-methylbenzenebutanoate, a selective, orally active leukotriene receptor antagonist. Can J Physiol Pharmacol 1986;64:1068–1075.
67 Fleisch JH, Rinkema LE, Haisch KD, et al: LY-171883, 1-[2-hydroxy-3-propyl-4-[4-(1H-tetrazol-5-yl)butoxyl]phenyl]ethanone, an orally active leukotriene D_4 antagonist. J Pharmacol Exp Ther 1985;233:148–157.
68 Gleason JG, Hall RF, Perchonock CD, et al: High affinity leukotriene receptor antagonists. Synthesis and pharmacological characterization of 2-hydroxy-3-[2-carboxy-ethylthio]3-[2-(8-phenyloctyl)phenyl] propanoic acid. J Med Chem 1987;30: 959–961.
69 Barnes N, Evans J, Zakrzewski J, et al: Pharmacology and physiology of leukotrienes and their antagonists; in Roberto L, Krell RD (eds): Biology of the Leukotrienes. New York, New York Academy of Sciences, 1988, pp 369–378.
70 Britton JR, Hanley SP, Tattersfield AE: The effect of an oral leukotriene D_4 antagonist L-649,923 on the response to inhaled antigen in asthma. J Allergy Clin Immunol 1987;79:811–816.
71 Jones TR, Zamboni R, Belley M, et al: Pharmacology of L-660,711: a novel, potent and selective leukotreine D_4 receptor antagonist. Can J Physiol Pharmacol, in press.
72 Synder DW, Giles RE, Keith RA, et al: In vivo pharmacology of ICI 198,615: a novel, potent and selective peptide leukotriene antagonist. J Pharmacol Exp Ther 1987;243:548–556.

73 Gillard J, Ford-Hutchinson AW, Chan C-C, et al: L-663,536 (3-[3-(4-chlorobenzyl)-3-t-butyl-thio-5-isopropylindol-2-yl]-2,2-dimethylpropanoic acid): a novel, orally active leukotriene biosynthesis inhibitor. Can J Physiol Pharmacol, in press.
74 Woodward DF, Nieves AL, Gary RK, et al: Studies with a potent leukotriene-antagonist. SK&F 104353, reveal that peptidoleukotrienes play a major role in mediating experimental allergic conjunctivitis. Invest Ophthalmol Vis Sci 1987;28:200.
75 Hay DWP, Muccitelli RM, Tucker SS, et al: Pharmacologic profile of SK&F 104353: a novel, potent and selective peptidoleukotriene receptor antagonist in guinea pig and human airways. J Pharmacol Exp Ther 1987;243:474–481.
76 Chan C-C, Dagenais F, Firby P, et al: Immediate hypersensitivity reactions in the guinea-pig conjunctiva: studies with a second-generation LTD_4, receptor antagonist, MK-571. Can J Physiol Pharmacol, in press.
77 Chiou LY, Chiou GCY: Occular anti-inflammatory action of a lipoxygenase inhibitor in the rabbit. J Ocul Pharmacol 1985;1:383–390.
78 Verbey NLJ, van Haeringen NJ, de Jong PTVM: Modulation of immunogenic keratitis in rabbits by topical administration of inhibitors of lipoxygenase and cyclooxygenase. Curr Eye Res 1988;7:361–368.
79 Bhattacherjee P, Williams RN, Eakins KE: A comparison of the ocular anti-inflammatory activity of steroidal and nonsteroidal compounds in the rat. Invest Ophthalmol Vis Sci 1983;24:1143–1146.
80 Rao NA, Patchett R, Fernandez MA, et al: Treatment of experimental granulomatous uveitis by lipoxygenase and cyclo-oxygenase inhibitors. Arch Ophthalmol 1987;105:413–415.
81 Miyano K, Chiou GCY: Pharmacological prevention of ocular inflammation induced by lens proteins. Ophthalmic Res 1984;16:256–263.
82 Ben-Zvi A, Rodrigues MM, Gery I, et al: Induction of ocular inflammation by synthetic mediators. Arch Ophthalmol 1981;99:1436–1444.
83 Johnson HG: Cromoglycate and other inhibitors of mediator release; in Middleton E, Reed CE, Ellis EF (eds): Allergy. Principles and Practice. St. Louis, Mosby, 1983, vol 1, pp 613–632.
84 Friedlaender MH: Ocular allergy. Int Ophthalmol Clin 1985;25:35–41.
85 Manabe R, Matsuda H, Sakuragi S, et al: Clinical effect of 0.25% AA-673 ophthalmic solution on allergic and vernal conjunctivitis. Folia Ophthalmol Jap 1987;38:1108–1119.
86 Saijo T, Kuriki H, Ashida Y, et al: Mechanism of the action of amoxanox (AA-673), an orally active antiallergic agent. Int Arch Allergy Appl Immunol 1985;78:43–50.
87 Kuriki H, Ashida Y, Saijo T, et al: Effects of amlexanox and various antiallergic agents on histamine release from rat peritoneal mast cells and on leukotriene generation in guinea pig lung fragments. Jap Pharmacol Ther 1988;16:111–117.
88 Fujisaki Y: Clinical evaluation of AA-673 (amlexanox) tablets in nasal allergy. Ther Res 1988;8:335–345.

Chi-Chung Chan, PhD, Department of Pharmacology,
Merck Frosst Center for Therapeutic Research, CP/PO 1005,
Pointe-Claire-Dorval, Que H9R 4P8 (Canada)

Bazan NG (ed): Lipid Mediators in Eye Inflammation.
New Trends Lipid Mediators Res. Basel, Karger, 1990, vol 5, pp 56–64

Anterior Segment Inflammation and Trabecular Meshwork Cell Responses[1]

Jon R. Polansky[a], *Robert N. Weinreb*[b]

Department of Ophthalmology, University of California, [a]San Francisco and [b]San Diego, Calif., USA

Anterior segment inflammation is likely to result in the release of a number of cellular mediators as well as potentially toxic materials, including reactive species of oxygen and debris, which could influence trabecular meshwork function. Breakdown of the blood-aqueous barrier, influx of activated inflammatory cells, and direct injury to ocular tissues are all potential contributors to alterations in trabecular meshwork cell physiology. A substantial rise in intraocular pressure (IOP) and increased resistance to aqueous humor outflow are observed in some cases; this 'inflammatory glaucoma' may involve direct injury to trabecular cells and/or an inability of the cells' homeostatic mechanisms to overcome the obstruction.

Evaluations of trabecular meshwork cells in vitro have helped to demonstrate important properties and responses of this unique cell type which may contribute to the maintenance of the conventional aqueous outflow pathway; these properties include phagocytosis of debris, maintenance of an endothelial-like lining and structure of the intertrabecular spaces, and the synthesis of t-PA and other enzymes which help to remove obstructive materials [for review, see 1]. Changes in these and other trabecular cell functions may be 'physiological' when they act to preserve aqueous outflow channels, or 'pathological' when they are part of a process leading to elevated IOP. The rise of IOP associated with anterior uveitis is often improved by controlling the inflammation in the eye, but the pathogenesis of this condition has not been fully explained and requires further investigation.

[1] Supported by NIH grants EY-02477 (J.R.P.) and EY-05990 (R.N.W.).

In the following review, we summarize our work on human and monkey trabecular meshwork cell cultures and some potentially relevant responses of trabecular cells to factors which may be released into the aqueous humor during inflammation.

Review of Current Observations

As shown in figure 1, it is possible to propagate trabecular meshwork cells in serial culture as a single layer of endothelial-like cells which resemble the cells as they exist in the trabecular outflow pathway. Under the conditions we have described, trabecular cells can be generated from both human and primate sources for multiple passages at high split ratios which retain the important structural and functional properties of this distinctive cell type for experimental studies [e.g., see 1–5].

The ability of trabecular cells to produce prostaglandins (PGs) and other eicosanoids has provided one major approach to our research efforts because of the potential importance of these lipid mediators as regulators in the meswhork and other tissues involved in aqueous outflow (e.g., ciliary muscle cells and uveoscleral outflow). Changes in the production or release of prostaglandins in response to different factors could be relevant to the effects of inflammatory products on the outflow pathway.

Our initial findings emphasized that trabecular cells under both serum-stimulated and serum-free conditions produce a wide variety of arachidonic acid metabolites which could act as local regulators (autacoids) in the outflow pathway. Trabecular cells produce PGE_2 as their major eicosanoid with somewhat smaller amounts of $PGF_{2\alpha}$ and 6-keto-$F_{1\alpha}$, and a variety of putative lipoxygenase products [6–9]. Interestingly, the production of PGs appears to be substantially greater in the trabecular cell compared to the endothelial lining cells of the vasculature.

As shown in figure 2a, b, tumor necrosis factor (TNF) and interleukin-1 (IL-1) produce a dose-response increase in the production of PGE_2 by monkey trabecular meshwork cells. Since both of these factors can be released by inflammatory cells present in the anterior segment during inflammation, the observed effects may be relevant to the physiological responses of the trabecular cell in uveitis.

Table 1 shows the time-dependent effects of other factors which evaluated as putative regulators of trabecular cell PG production. Using serum-deprived human trabecular cell cultures, we observed that throm-

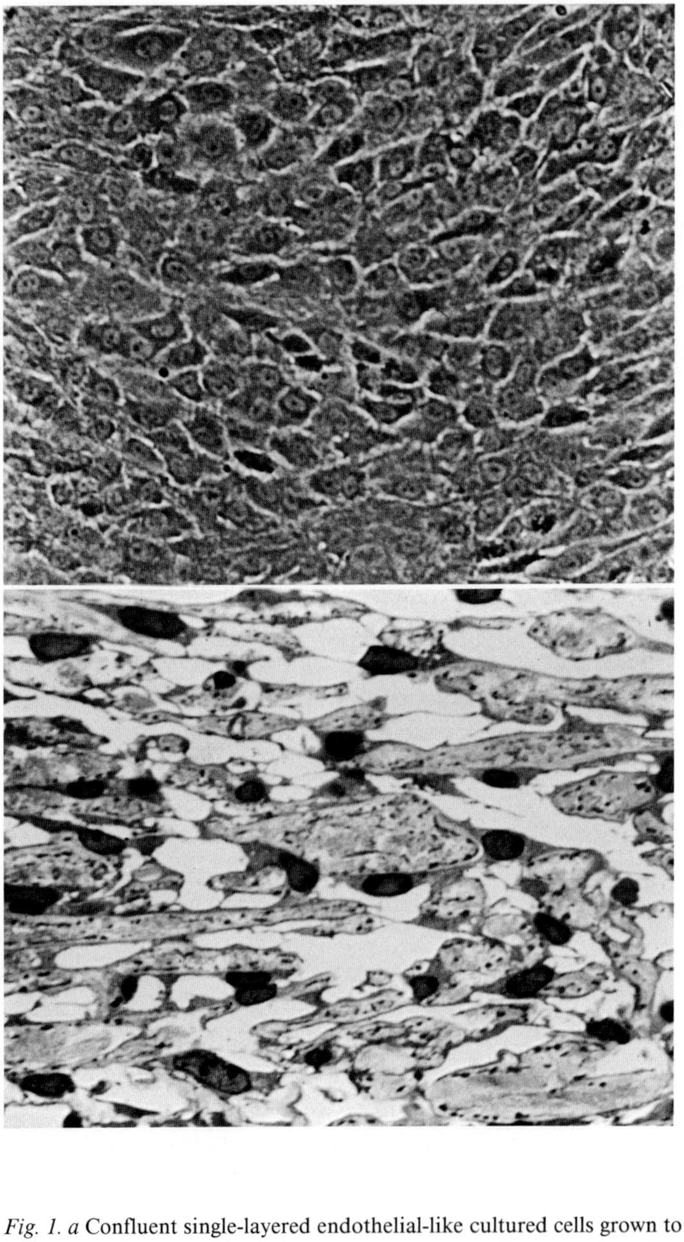

a

b

Fig. 1. a Confluent single-layered endothelial-like cultured cells grown to the fourth passage (top), and human trabecular meshwork cells. *b* Cross section of corneoscleral meshwork tissue (bottom) demonstrating the endothelial-like lining properties [modified from 1].

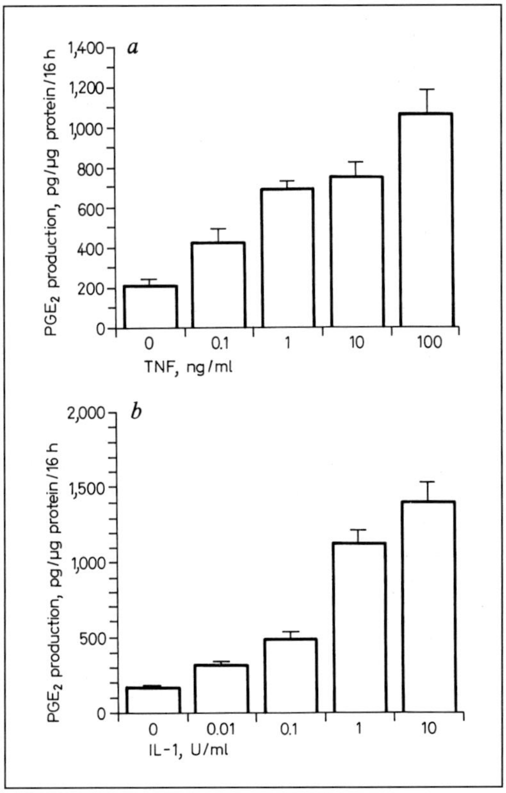

Fig. 2. Monkey trabecular meshwork cells: dose-response effects of (*a*) TNF and (*b*) IL-1. Trabecular meshwork tissue was obtained from a 3.2-kg cynomolgus monkey and propagated in serial culture with 10% FCS [5]. Fourth and fifth passage cells were employed for these studies, in which varying concentrations of human IL-1 (Genzyme, Boston, Mass.) at 0.01, 0.1, 1.0 and 10 U/ml, or TNF at 0.1, 1.0, 10 and 100 ng/ml were added to the culture media. Control dishes received no additions. Incubations were conducted for 16 h, media collected and cells harvested. All experiments were performed in quadruplicate [from Weinreb and Mitchell, unpubl. data].

bin, platelet-activating factor (PAF), and serum (constituents which can be produced during anterior segment inflammation and breakdown of the blood-aqueous barrier) produced stimulation of PGE_2 levels. The rapid and large response of trabecular cells to bradykinin and thrombin were notable. The delay in the response to PAF and fetal calf serum (FCS) requires further explanation, but suggests that different mechanisms may

Table 1. Influence of stimulators of human trabecular cell PGE_2 production: time dependence of the response

	1 h	3 h
Bradykinin	4 +	4 +
Thrombin	4 +	4 +
PAF	1 +	3 +
FCS	1 +	5 +
CI	1 +	1 +
Control	0	0

Confluent fourth passage human trabecular cells from a 30-year-old patient were incubated with bradykinin (10^{-5} *M*), thrombin (1 U/ml), PAF (10^{-9} *M*), calcium ionophore A23187 (CI, 10^{-5} *M*), or FCS (10%). The cultures were changed into DMEM with no FCS 1 h prior to the experiment. PGE_2 levels in the treated versus control groups were determined at different times after addition of the individual stimuli. Responses were graded according to a relative scale (1+ = 0.5- to 1.5-fold increase, 4+ = 6- to 10-fold increase, 5+ = 10-fold increase, with 2+ and 3+ showing intermediate values). Basal levels ranged from 2 to 10 ng/ml PGE_2 in different experiments for the 1- and 3-hour time-points [from 9].

be involved with the responses to these agents. The relatively smaller effect of calcium ionophore, employed as a 'positive control' in the experiment, may involve an effect of this agent to produce cellular damage in addition to opening calcium channels. As described below, membrane damage to human trabecular cells produces a decrease in PG production, perhaps accounting for a smaller stimulatory effect of the ionophore.

The interaction of oxidation products with living cells damages membranes and often produces a *stimulation* of PGs as measured in vascular cells, lens cells, and other cell types, which may be involved in the mechanism of cell injury. Membrane damage and other effects of hydrogen peroxide on human trabecular cells have been of interest to us in relation to primary open-angle glaucoma [e.g. see 1, 10] in addition to inflammatory glaucoma. Hydrogen peroxide and other reactive species of oxygen will be generated to a greater than normal degree by inflammatory processes in the anterior segment, and these substances are believed to be involved in the pathogenesis of cataract and in the damage to ocular and systemic tissues observed with anoxia and inflammation.

As shown in figure 3 and reported in our earlier studies [10], treatment of trabecular cells with both sublethal (0.6 m*M* bolus treatment) and lethal

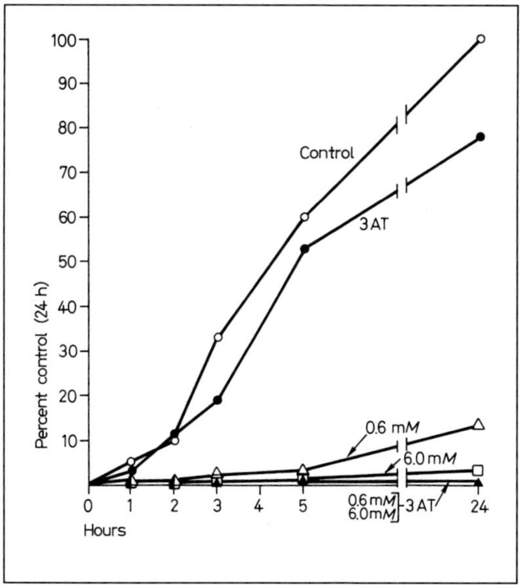

Fig. 3. Hydrogen peroxide effects on PG production by human trabecular cells. Fifth passage confluent trabecular cells from a 30-year-old subject were incubated in PBS medium with and without 3AT (30 m*M* catalase inhibitor) for 1 h before addition of 0.6 or 6.0 m*M* H_2O_2, which was incubated with the cells for an additional hour [Barksdale and Polansky, unpubl. observations]. PGE_2 levels were determined at the times indicated up to 24 h after the medium was changed and the cells placed back into normal culture conditions (DMEM with 10% FCS).

(6 or 0.6 m*M* with 3-aminotriazole (3AT) catalase inhibitor) concentrations of H_2O_2 produced a marked *decrease* in PGE_2 production. A dose-dependent decrease in PG production was observed beginning with concentrations only slightly above that believed to be present physiologically in the aqueous humor (fig. 4). Peroxide concentrations greater than 0.06 m*M* resulted in a substantial decrease in trabecular cell PGE_2 production, with a greater than 95% maximal suppression occurring before changes in cellular morphology were observed. Morphological alterations were not apparent until 0.6–1 m*M* H_2O_2, and cell death did not occur until higher doses were administered. Controls performed in our laboratory under identical experimental conditions using human lens cells and vascular endothelial cells showed the expected response dose-dependent increase in PG production.

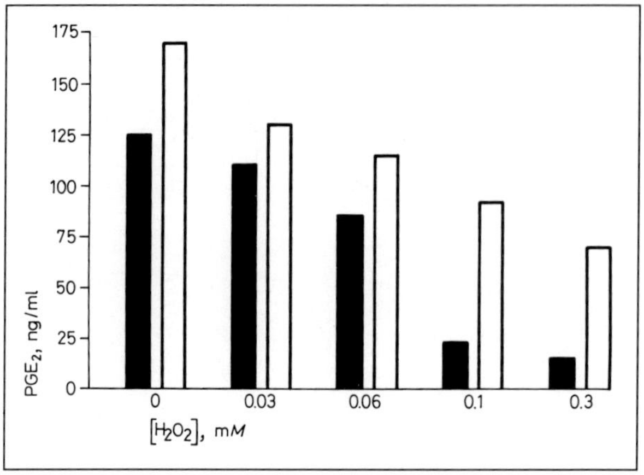

Fig. 4. Peroxide inhibition of PG production by human trabecular cells: dose-response study. Confluent fifth passage cells from 14- (■) and 30-year-old (□) subjects were incubated with varying concentrations of H_2O_2 for 1 h in PBS medium, and PGE_2 measured at 24 h after return of the cells to DMEM with 10% FCS [Fauss and Polansky, unpubl. observations].

The decrease in PG production by trabecular meshwork cells may signal a shutting down of important cellular functions involving membrane damage and other alterations, or may be part of a yet to be determined homeostatic mechanism since the change is opposite that of the other cell types examined. Trabecular cells do appear to have a large capacity to detoxify H_2O_2 from their environment, and it will be useful to learn more concerning the defense mechanisms by which trabecular cells maintain themselves in the face of increased oxidative and other stresses during inflammation. It will also be important to know whether other reactive species of oxygen (e.g. lipid oxides) may be more damaging to trabecular cells than H_2O_2.

Comment

Inflammatory processes involving the anterior segment may produce alterations in the IOP. There may be an associated hypotony due to decreased aqueous humor secretion and/or an increase of the uveoscleral pathway. IOP may also increase due to the development of inflammatory glaucoma. In the latter case, the trabecular meshwork may become occluded by peripheral anterior synechiae, or there may be a wide open angle with a possible direct effect on trabecular meshwork function.

We have also been evaluating experimental animal systems [e.g., see 12, 13] to learn more about changes which can take place in the physiology of the anterior segment with injury and inflammation. In vitro approaches complement such studies and permit evaluations to be conducted on individual ocular cell types in which specific inflammatory factors and potentially injurious chemicals can be examined. In vivo correlations may help establish links with clinically relevant conditions.

If PGs play a role in mediating outflow facility and/or trabecular cell function, the direct effects of inflammatory mediators demonstrated in our studies may facilitate aqueous outflow directly, or play a role in other processes which preserve open channels. A decreased amount of PGs produced by trabecular cells following injury due to increased levels of H_2O_2 (or other reactive species of oxygen) may participate in the unexplained rise of IOP in certain cases of uveitis – although this proposal requires considerably more investigation.

Alterations in other eicosanoids in addition to PGs (e.g. products of lipoxygenase or P-450 pathways) and other lipid mediators (e.g. inositol phosphates), as well as effects on specific trabecular cell functions (e.g. phagocytosis, connective tissue production, and integrity of the cellular lining) are currently being evaluated. Hopefully, such approaches will yield additional information to better understand changes in the aqueous outflow pathway in the presence of anterior segment inflammation.

References

1 Polansky JR, Wood IS, Maglio MT, et al: Trabecular meshwork cell culture in glaucoma research: evaluation of biological activity and structural properties of human trabecular cells in vitro. Ophthalmology 1984;91:580–595.

2 Polansky JR, Weinreb RN, Baxter JD, et al: Human trabecular cells. I. Establishment in tissue culture and growth characteristics. Invest Ophthalmol Vis Sci 1979; 18:1043–1049.

3 Alvarado JA, Wood I; Polansky JR: Human trabecular cells. II. Growth pattern and ultrastructural characteristics. Invest Ophthalmol Vis Sci 1982;23:464–478.

4 Polansky JR, Weinreb R, Alvarado JA: Studies on human trabecular cells propagated in vitro. Vision Res 1981;21:155–160.

5 Ryder MI, Weinreb RN: The cytoskeleton of the cynomolgus monkey trabecular cells. I. General considerations. Invest Ophthalmol Vis Sci 1986;27:1305–1311.

6 Weinreb RN, Mitchell MD, Polansky JR: Prostaglandin production by human trabecular cells in in vitro inhibition by dexamethasone. Invest. Ophthalmol Vis Sci 1983;24:1541–1545.

7 Weinreb RN, Polansky JR, Alvarado JA, et al: Arachidonic acid metabolism in human trabecular meshwork cells. Invest Ophthalmol Vis Sci 1988;29:1708–1712.

8 Weinreb RN, Mitchell MD: Prostaglandin production by cultured cynomolgus monkey trabecular meshwork cells. Prostaglandins Leukotrienes Essent Fatty Acids, in press.

9 Polansky JR, Kurtz RM, Alvarado JA, et al: Eicosanoid production and glucocorticoid regulatory mechanism in cultured human trabecular cells; in Bitzo LZ, Stjernschantz J (eds): The Ocular Effects of Prostaglandins and Other Eicosanoids. New York, Liss, 1989.

10 Barksdale S, Ghuan K, Podos S, et al: Oxidative stress in cultured human trabecular cells: Suppression of prostaglandin production. Invest Ophthalmol Vis Sci 1986; 27(suppl):210.

11 Mitchell MD, Weinreb RN: Modulation of arachidonic acid metabolism by substances in aqueous humor. Prostaglandins Leukotrienes Med 1985;18:309–311.

12 Peyman G, Fishman PH, Alexander KR, et al. The effect of ultraviolet, visible, and infrared irradiation on the rabbit blood-aqueous barrier. Exp Eye Res 1986;42:249–254.

13 Weinreb RN, Weaver D, Mitchell MD: Prostanoids in rabbit aqueous humor: effect of laser photocoagulation of the rabbit iris. Invest Ophthalmol Vis Sci 1985;26: 1087–1092.

Jon R. Polansky, MD, Cellular Pharmacology Laboratory, Room K301,
UCSF, Box 0730, San Francisco, CA 94143 (USA)

Bazan NG (ed): Lipid Mediators in Eye Inflammation.
New Trends Lipid Mediators Res. Basel, Karger, 1990, vol 5, pp 65–82

Inflammatory Mediators in Models of Immunogenic and Nonimmunogenic Inflammation of the Anterior Segment of the Eye[1]

P. Bhattacherjee, Christopher A. Paterson[2]

Department of Ophthalmology and Visual Sciences, Kentucky Lions Eye Research Institute, University of Louisville, School of Medicine, Louisville, Ky., USA

Inflammation is a defensive response of tissue to trauma, foreign materials or immunocomplexes. Ocular inflammation is characterized by vasodilation, disruption of the blood-aqueous barrier and cellular infiltration of ocular tissues and fluids. Inflammatory responses of a mild or moderate nature are self-limiting; in severe inflammation, the functional integrity of the visual system may be threatened. In severe inflammation, appropriate drug therapy, or even surgical intervention, becomes essential.

In the last few decades considerable advances have been made in elucidating the role of chemical mediators and immune mechanisms in inflammation. The actions and interactions of known chemical mediators such as prostaglandins (PGs), leukotrienes and related eicosanoids, which are elaborated by the tissues and the invading inflammatory cells, have been extensively studied. These metabolites of arachidonic acid have a vast array of biological actions ranging from vasodilation and chemotaxis to immunomodulation. A detailed discussion of the biosynthesis of PGs and leukotrienes from arachidonic acid by the ocular tissues, and the inflammatory actions of these substances, is provided in another chapter in this book [Kulkarni, 'Eicosanoids in Ocular Inflammation']. In addition to the arachidonic acid derived inflammatory mediators, recently discovered substances like interleukin-1 (IL-1) and platelet-activating factor

[1] Preparation of this chapter was supported in part by USPHS research grant No. EY06918, the Kentucky Lions Eye Foundation, and an unrestricted award from Research to Prevent Blindness, Inc.

[2] We thank Elizabeth Paterson for excellent editorial assistance.

(PAF) have inflammatory properties. Furthermore, these same agents can elicit the release of arachidonic acid metabolites from various tissues and inflammatory cells. The actions and interactions of these and other mediators during inflammatory reactions and activation of the immune component, remain to be fully appreciated.

Despite considerable progress in inflammation research, our understanding of inflammatory diseases is limited. The quest to understand the etiology and mechanisms of ocular inflammatory diseases is hampered by the lack of entirely suitable animal models of ocular inflammation. The utility of animal models is compromised because the pattern and action of inflammatory mediators produced by animal tissues are not always similar to those in humans. Some of the inflammatory mediators elaborated by animal species are not formed by human tissues and the actions of the mediators differ in nature and magnitude. However, in the absence of satisfactory alternatives to animal models, we must continue to use this approach to gather information that can be cautiously extrapolated to human disease.

The methodology to induce experimental ocular inflammation is very diverse. Some of the inflammatory stimuli that have been employed were used to explore quite specific aspects of ocular inflammation. For instance, models of immunogenic anterior uveitis and autoimmune uveoretinitis were developed primarily to study the clinical characteristics of inflammatory reactions and the role of the immune system. On the other hand, several animal models of nonimmunogenic uveitis were developed with the purpose of investigating the elaboration, release and action of putative or established chemical mediators of inflammation.

In this chapter, selected models of ocular inflammation will be discussed in terms of their utility and the involvement of lipid mediators in the inflammatory response. The purpose of the chapter is not to present an exhaustive review but rather to illustrate the value of an animal model of ocular inflammation to assist in unravelling the complex topic of inflammatory mediators.

Neurogenic Inflammation

Injury of the eye is universally associated with pain. For some time it has been known that sensory fibers of the fifth nerve are involved in pupil constriction and corneal sensation [1, 2]. The ocular responses of the rabbit eye to sensory nerve stimulation were first described in detail by Perkins [3]. This study demonstrated that electrical stimulation of the fifth nerve causes pupil constriction, breakdown of the blood-aqueous barrier, and the rise in intraocular pressure (IOP). Almost 20 years after Perkins' observations, a few studies examined whether or not the responses to substances such as nitrogen mustard, formaldehyde and laser irradiation are neurogenic [4–6]. Indeed, these studies reported similarities of the responses to those elicited by fifth nerve stimulation. Formaldehyde-induced responses are of short duration, lasting for about 60 min [5]. In contrast, nitrogen mustard induces a biphasic response; the first phase consisting of miosis, IOP rise and protein exudation is neurogenic and lasts for 60 min. During the second phase, aqueous humor protein content is further increased with accompanying leukocyte infiltration between 3 and 12 h; the responses of the second phase are thought to be mediated by prostaglandins [6]. The work of Jampol and Noth [7], Butler et al. [5] and Camras and Bito [6] demonstrated that the response to formaldehyde and the first phase of the response to nitrogen mustard are not inhibited by inhibitors of the cyclooxygenase pathway of arachidonic acid metabolism (leading to PGs) but are abolished by sensory denervation; thus it appears that these responses are neurogenic. The inflammatory response elicited by direct trigeminal nerve stimulation might be a cleaner model of neurogenic inflammation because it avoids nonspecific effects of chemical irritants such as tissue damage, which might elicit cellular infiltration.

The nature of chemical mediators released from the sensory neurons during chemical stimulation is not fully known. Since 1931, substance P (SP) was known as a noncholinergic smooth muscle contracting agent [8]. In 1953, SP was suggested to be a sensory transmitter after it was found in dorsal roots [9, 10]. The intensive exploration of SP as a sensory nerve mediator began only after delineation of the structure and synthesis [11, 12]. A concise history of SP is given by Otsuka and Konishi [13]. Olgart et al. [14] observed that SP is released during the stimulation of the mandibular nerve. The release of SP in the rabbit eye during stimulation of the fifth nerve has also been demonstrated [15, 16]; the source of SP detected in the anterior chamber appears to be sensory nerves [17]. There is histochemical and biochemical evidence that SP is present in the sensory nerves of animal and human ocular tissues [18–21]. Synthetic SP given intraocularly is a potent miotic [17, 22–24]; its role in the disruption of the blood-aqueous barrier or rise in IOP is questionable. Some studies have reported effects of SP on the blood-aqueous barrier and IOP [22, 23] while others found SP without an effect on these parameters [16, 25, 26, and Bhattacherjee, unpubl. data]. Therefore, SP may not be involved in the neurogenic responses other than miosis. Probably, other peptides such as vasoactive intestinal peptide (VIP) and calcitonin-gene-related peptide (CGRP) located in the sensory and sympathetic neurons also participate in neurogenic responses. There are reports that VIP and CGRP cause neurogenic plasma extravasation and IOP rise and cellular infiltration [27–29]. More studies are needed to elucidate the roles of various neuropeptides in neurogenic inflammation.

Allergic Conjunctivitis

The most predominant inflammatory reactions seen in the conjunctiva are of the allergic type. The models that have been developed are of immediate and delayed hypersensitivity reactions.

In 1965, Aronson et al. [30] described a model of conjunctival inflammation in rabbits immunized with bovine serum albumin (BSA) given topically to the conjunctiva for several weeks. Conjunctival follicles, limbal hyperemia and infiltration by plasma cells and small lymphocytes developed after continuous immunization for 6 weeks. This model was slightly modified by Hall and Pribnow [31], who used a more vigorous topical treatment of the conjunctiva with ovalbumin, thus shortening the immunization period.

In the model of Dwyer et al. [32], allergic conjunctivitis was induced in the sensitized guinea pigs by multiple topical administration of serum albumin. The procedure involves sensitization of the animals with intradermal injections of 0.1 ml of rabbit serum albumin at day 1 and 10. The conjunctiva of the sensitized animal is then challenged by topical administration of the same antigen. Within a few minutes of the challenge, vascular congestion and edema are evident. In the next 4–24 h tissues become infiltrated by eosinophils and neutrophils; these reactions are thought to be IgE mediated. The inflammatory responses in this guinea pig model were assessed qualitatively. However, the degree of formation of edema (extravasation of plasma protein) could be readily quantified by measuring the radioactivity in the conjunctival tissues following an intravenous injection of ^{125}I-labelled albumin prior to sacrifice of the animals. The degree of leukocyte infiltration was determined semiquantitatively from the histological slides. Alternatively, PMN infiltration could be quantified by measuring myeloperoxidase (MPO) in the conjunctival tissue as described by Williams et al. [33].

Delayed hypersensitivity reactions have been induced in the conjunctiva of guinea pigs sensitized with intradermal injection of oxazolone with either the Freund's incomplete or complete adjuvant [34]. Twelve hours after the challenge conjunctival vessels become congested and at 24 h mononuclear cells and eosinophils infiltrate the tissues. The severity of these responses is greatest when the antigen is given together with the Freund's complete adjuvant.

In the models of conjunctivitis described above, the nature of the chemical mediators elaborated and released after the inflammatory stimuli has not been extensively studied. There are, however, investigations which have examined the conjunctival responses to known chemical mediators such as histamine, PGs, leukotrienes and PAF. The conjunctiva contains large numbers of mast cells; following the induction of uveitis by BSA, degranulating mast cells were observed in the conjunctival tissues [35]. Allansmith et al. [36] made similar observations in immunized rats undergoing immediate hypersensitivity reactions following a systemic challenge. Thus, histamine released from the mast cells probably plays only some complementary roles since antihistamines and sodium cromoglycate, a mast cell stabilizer, have limited beneficial effect in allergic conjunctivitis [37].

Eakins and Bhattacherjee [38, 39] demonstrated that histamine given topically, alone, was only mildly vasodilatory in the rabbits but when given together with PGs, conjunctival edema and leukocyte infiltration occurred. The implication of these observations is that histamine must act cooperatively with other inflammatory mediators to induce the full picture of allergic conjunctivitis. Conjunctival reactions to synthetic leukotrienes given topically to the conjunctiva have been examined in rabbits and guinea pigs. LTB_4 causes PMN infiltration in the rabbit conjunctiva [40], but in the guinea pig conjunctiva, large doses are needed [41]. Peptidoleukotrienes, LTD_4 and LTE_4, but not LTC_4, cause eosinophilic infiltration in the guinea pig after topical administration [41]. In addition to the chemotactic activity of leukotrienes, LTD_4 in particular has been shown to increase microvascular permeability [42, 43] and potentiate chemotactic activity of LTB_4 [44]. As noted earlier, no studies have been undertaken to examine the release of leukotrienes into the conjunctival exudate or the role of leukotrienes in the experimental model of allergic conjunctivitis.

However, one study in humans [45] reported the presence of LTD_4 in the tear fluid after antigen challenge of the conjunctiva, indicating a potential role of leukotrienes in allergic conjunctivitis.

It is evident that the role of lipid mediators in allergic reactions in the conjunctiva is poorly understood. The animal models that are available would seem very adequate to explore this facet of the inflammatory response.

Corneal Inflammation

There is a wide variety of animal models of corneal inflammation, the majority of which are nonimmunogenic. With its epithelium intact, the cornea is not readily accessible to microbial pathogens and large molecules. When the corneal epithelium is damaged or mechanically removed, the cornea becomes very permeable to pathogens. Blood-borne elements also have access to the cornea by way of the limbal circulation. For simplicity, this section will be divided into models of corneal inflammation that are immunogenic and nonimmunogenic.

Immunogenic Corneal Inflammation

Immunogenic inflammation of the cornea can be induced in sensitized animals by an intracorneal challenge with a sensitizing antigen [46]. Alternatively, a delayed type of immune response can be elicited in rabbits or guinea pigs following direct injections of antigen into the cornea [47, 48]. The reactions to antigen injection are biphasic, the first phase consisting of corneal opacity and infiltration of the limbal region by plasma cells and lymphocyte occurring between 3 and 5 days. This is followed by the second phase, between 7 and 14 days, consisting of corneal deposition of immune complexes, infiltration by plasma cells. These responses are enhanced if the antigen is injected together with Freund's complete adjuvant [49].

The chemical mediators involved in immunogenic corneal inflammation have not been examined. However, the release of arachidonic acid metabolites during corneal inflammation is expected, since corneal tissues have the ability to metabolize arachidonic acid. There are reports that corneal epithelial cell culture, or an established cell line, forms a substance with biological properties similar to IL-1 [50, 51]. Further, recombinant IL-1 incorporated in Elvax pellets and implanted into the cornea causes neovascularization, a feature occasionally accompanying keratitis. Thus, it can be speculated that IL-1 is also participating in corneal inflammation.

Nonimmunogenic Corneal Inflammation

In view of the many approaches to initiating an inflammatory response in the cornea, we will discuss only a limited number including intrastromal clove oil injection, alkali burn, cryosurgery and *Pseudomonas aeruginosa* infection. Generally speaking, it is difficult to incite an inflammatory response that is truly limited to the cornea. As a result of a neurogenic response and the close proximity of all other anterior segment tissues, most models of corneal inflammation will involve at least a mild anterior uveitis.

The introduction of microliter amounts of clove oil into the rabbit cornea induces a reproducible inflammatory response [52, 53]. Although the specific mediators have not been detailed, the responsiveness of the inflammation to steroid therapy [54] would perhaps suggest the involvement of mediators derived from arachidonic acid metabolism. Another approach to eliciting a localized corneal inflammatory response is by applying a cryoprobe to the corneal surface [55] or by scraping away a portion of the corneal epithelium [56]. Bazan [55] showed that the cryoinjured cornea generates several products of both the cyclooxygenase and lipoxygenase pathways of arachidonic acid metabolism. The inflammatory mediators released following corneal scraping have not been specifically identified.

A massive anterior segment inflammatory response together with epithelial or penetrating stromal ulcers can be induced by intrastromal injection of microbial pathogens [e.g. 57]. The pathological responses of the cornea to the intrastromal injection of *P. aeruginosa* in experimental animals is analogous to infectious keratitis of the human cornea; the response consists of the development of a white opaque stromal ring within 24 h, which is associated with leukocyte infiltration, conjunctival hyperemia and haziness of the cornea by 48 h. Stromal melting, ulceration, descemetocele formation, and corneal perforation can proceed rapidly. The corneal pathology is most likely caused by proteases including elastase released from *P. aeruginosa,* and host-derived collagenase from invading PMNs [57–59].

Corneal inflammation and ulceration induced by corneal alkali injuries [60–62] has drawn wide attention in recent years. By varying the duration of the exposure, the strength of the alkali (usually sodium hydroxide) and the area of cornea burned, mild to severe corneal ulceration and perforation can be obtained in rabbit eyes. Alkali burns (1 *N* sodium hydroxide) of 6 mm of corneal surface cause epithelial defects but do not produce ulceration. Persistent epithelial erosion and ulceration progressing to perforation occurs after alkali injury to a much larger corneal surface area. The inflammatory reactions that follow alkali treatment are not confined to cornea only but also involve the entire anterior segment.

The specific mediators involved in the corneal inflammation remain to be fully elucidated. Normal cornea is capable of synthesizing PGs and other products of arachidonic acid metabolism [for details, see Kulkarni, this volume]. Bovine and rabbit corneal epithelium can metabolize arachidonic acid also via cytochrome P-450 pathway to 12(R)-hydroxy-5,8,10,14- and 12-hydroxy-5,8,14-eicosatetraenoic acid [63]; the latter metabolite in 5–10 ng doses causes vasodilation and disruption of the blood-aqueous barrier in rabbits [64]. It is not yet known whether any of these corneal P-450-derived substances have a role in corneal inflammation.

Another recently discovered putative inflammatory mediator, PAF, shares the inflammatory actions of arachidonic acid metabolites. To date there is only one study showing cornea is able to synthesize PAF after alkali burn [65]. Another study (cited by Braquet et al. [66]) showed that a PAF antagonist alleviates keratitis, suggesting involvement of PAF in corneal inflammation.

Uveitis

The uveal tract, comprised of iris, ciliary body and choroid, is highly vascularized, and therefore very vulnerable to inflammatory disease. The etiology of uveitis is very diverse, including infection, allergy, trauma and noninfectious systemic disease. Uveitis developing as a result of infection is usually acute, but in circumstances where cellular and humoral mechanisms do not contain the response, the chronic phenomena of immunological hypersensitivity develop. The uveitis associated with systemic disease such as ankylosing spondylitis, Reiter's syndrome, sarcoidosis, Behçet's disease and juvenile rheumatoid arthritis does not lend itself well to the development of animal models. The focus of this section will be placed upon those models which fall into the category of immunogenic or nonimmunogenic.

Immunogenic Uveitis

Active and passive immunogenic inflammatory reactions in the eye are induced by direct intraocular injection of the antigen in sensitized animals or antigen-antibody complex in nonsensitized animals. Eakins et al. [67] produced acute anterior uveitis in rabbits by injecting BSA into the vitreous body. The onset of the inflammatory reactions in this model varies from 8 to 18 days after BSA; the vascular and cellular inflammatory reactions last for about 10 days. An alternative procedure to induce similar immunogenic ocular inflammation is to sensitize either the rat or rabbit by subcutaneous injection of BSA or ovalbumin at 3-day or weekly intervals. One week after the last injection, intraocular inflammation is induced by an intravitreal challenge of the antigen (1 mg per eye) [68, 69]. The conjunctiva becomes hyperemic and edematous within 1 h of the challenge and a full inflammatory response develops by 24 h and lasts for 2–5 days. Polymorphonuclear leukocytes and monocytes are the predominant inflammatory cells present.

Intraocular tissues are highly vulnerable to autoimmune diseases. Sympathetic ophthalmia and Behçet's syndrome are examples of such diseases in humans. Several investigators have attempted to induce autoimmune uveitis with homologous or heterologous uveal or retinal antigen alone or together with Freund's complete adjuvant [70–75]. Wacker et al. [76] isolated and characterized a soluble antigen (S-antigen) from photoreceptors of guinea pigs, rabbits and bovine retina. Uveoretinitis in guinea pigs, rats and primates has been induced by subcutaneous injection of purified S-antigen in complete Freund's adjuvant [77–79]; inflammation of uvea and retina is progressive and dependent on antigen dose. An injection of 50 μg S-antigen results in severe inflammation of the intra- and extraocular tissues. The choroidal and retinal vasculature become necrotic with extensive infiltration by the inflammatory cells. Animals sensitized with smaller doses show different cellular characteristics. For example, 10 μg of S-antigen induces granulomatous infiltration with presence of lymphocytes and epithelioid cells. Epithelioid cells and giant cells are present between Bruch's membrane and retinal pigment epithelium. These inflammatory characteristics are comparable to those of sympathetic ophthalmia in humans [77]. Nussenblatt et al. [79] examined ocular inflammatory and immune responses in S-antigen-induced uveitis in rhesus monkeys. In addition to the inflammatory reaction in the anterior uvea, retinal and subretinal lesions, vasculitis and periphlebitis of the retinal vessels were observed. These pathological changes and reactivity of lymphocytes from S-antigen-sensitized monkeys were similar to human disease. Thus S-antigen-induced uveitis in monkeys is thought to be a good model of human uveitis.

One might expect the pattern of mediator release and activity to be most complex in immunogenic uveitis. As outlined in discussing conjunctivits, there is no doubt considerable interplay between lipid mediators and components of the immunological response. An increased level of PGE-like activity in the aqueous humor of rabbit eyes with BSA-induced uveitis has been detected [67]. In humans, no PGE-like activity was present in the aqueous humor of eyes with uveitis under steroid therapy, nor in eyes with cataract; PGs were detected in the aqueous humor of patients with acute uveitis, glaucomatocyclitis and Behçet's disease [80]. LTD_4 has also been detected in the aqueous humor of uveitis patients [81]. Whether the presence of these lipid mediators reflects direct or coincidental involvement in the disease process remains to be elucidated.

The role of IL-1 and PAF in experimental models of immunogenic uveitis is largely unknown. IL-1, a polypeptide elaborated by a variety of tissues and inflammatory cells, has diverse inflammatory and immunostimulant properties. It is highly chemotactic for leukocytes and degrades cartilage [82], releases metalloproteinases and arachidonic acid metabolites [83, 84] and activates lymphocytes [85]. Endowed with these wide-ranging activities, it will be surprising if IL-1 is not involved in immunogenic uveitis. There is only one study by Prendergast et al. [86] reporting the presence of IL-1-like activity in the aqueous humor of rabbit eyes with ovalbumin-induced uveitis.

Recombinant IL-1 injected into the anterior chamber of rabbits [87] resulted in the dose-dependent accumulation of PMNs in the anterior chamber with no effect on vascular permeability. The response was maximal at 4 h and persisted for 24 h. In contrast to these observations, Rosenbaum et al. [88] reported that intravitreal injection of IL-1 to rabbits caused the disruption of the blood-aqueous barrier, mononuclear and PMN cell infiltration lasting for 48–72 h. In view of the importance of IL-1 in inflammatory processes and immune system, its role in immune and autoimmunogenic uveitis needs in-depth exploration.

Nonimmunogenic Uveitis

Nonimmunogenic ocular inflammation of the anterior uvea can be induced in a number of animal species by intraocular, intravenous or footpad injection of phlogistic agents; most commonly used are gram-negative bacterial lipopolysaccharides (LPS). Intravenous injection of LPS into rabbits produces short-lasting, mild intraocular inflammation [89]. The intravitreal injection of LPS in rats or rabbits, on the other hand, induces moderate to severe intraocular inflammation [90–94]. The advantage of intravitreal injection is that the systemic reactions of intravenous injection of endotoxin are avoided.

Injections of 10–100 ng of LPS into the vitreous body of the rabbit eye dose-dependently cause vasodilation, disruption of the blood-aqueous barrier and leukocyte infiltration which are maximal at 24 h and persist for at least 72 h. Perhaps a more clinically relevant model is the one in which LPS is given into the footpad of rats. An injection of 100–300 μg/kg of LPS into the footpad produces conjunctival and uveal vascular and cellular responses lasting for about 72 h [95, 96]; the responses are biphasic. The first phase occurs around 4 h and consists principally of vasodilation and the disruption of the blood-aqueous barrier, resulting in increased plasma protein in the aqueous humor, without leukocyte infiltration. The second phase includes a further increase in aqueous humor protein coincident with massive leukocyte infiltration of the tissues and aqueous humor. The footpad injection of LPS also produces systemic reactions such as shock, and inflammation of the lungs and gastrointestinal tract persisting for less than 24 h.

The ocular inflammatory responses to LPS can be quantified by using a number of well-established methods [96]. The magnitude of disruption of the blood-aqueous barrier is usually quantified by measuring protein concentration in the aqueous humor. Alternatively, protein influx into the anterior chamber can be quantified by measuring radioactivity in the aqueous humor associated with ^{125}I-albumin injected intravenously immediately after inflammatory stimuli. Leukocyte accumulation in the anterior chamber is estimated by counting the number of leukocytes in the aqueous humor. A more quantitative measurement of PMN infiltration is the determination of MPO activity in the ocular tissues as reported by Williams et al. [33]. MPO being specific for PMNs, its measurement is preferred to visual counting of cells in the aqueous humor or in histological sections. The LPS-induced model of ocular inflammation in rats and rabbits is therefore very suitable for studying the mechanisms and factors involved in acute nonspecific intraocular inflammation.

There is considerable evidence to support the role of PGs and leukotrienes in nonimmunogenic ocular inflammation. PGs, particularly PGE_2, have been detected in the aqueous humor of rabbits and rats after intravitreal or intravenous injection of LPS [90, 93, 97, 98], and in more recent studies, the temporal relationship between PGE_2 and LTB_4 release and inflammatory reactions of the rabbit eye to intravitreal endotoxin have been examined [94, 99].

Pharmacological studies with the inhibitors of arachidonic acid metabolism or with anti-inflammatory steroids have provided additional support to the involvement of arachidonic acid metabolites in ocular inflammation. Most of these studies examined the effects of the compounds on the inflammatory responses, but not the formation or release of arachidonic acid derived mediators. Prednisolone or dexamethasone significantly reduce inflammatory reactions, particularly PMN infiltration in endotoxin-induced uveitis [100–102]. The anti-inflammatory actions of steroids are considered to be due to the inhibition of PLA_2, which permits the release of arachidonic acid. Dual cyclooxygenase and lipoxygenase inhibitors, or selective 5-lipoxygenase inhibitors, have been shown to reduce leukocyte infiltration in endotoxin-induced uveitis in rats and rabbits. Bhattacherjee et al. [101] studied the effects of BW755C, an inhibitor of cyclooxygenase and lipoxygenase enzymes, on the LPS-induced ocular inflammation in rats. This compound was found to significantly inhibit the leukocyte infiltration. Another compound, BWA4C, a highly selective lipoxygenase inhibitor, was reported to attenuate cellular infiltration and LTB_4 formation in the LPS-induced inflamed rabbit and rat eyes [99,103]. Thus, it can be further conjectured that arachidonic acid metabolites are involved in this model of intraocular inflammation.

There are a few studies reporting the inhibition of PGE_2 and LTB_4 release by cyclooxygenase and 5-lipoxygenase inhibitors. Bhattacherjee [90] reported that indomethacin reduced PGE_2 release with concomitant attenuation of protein influx into the anterior chamber. Herbort et al. [98] also demonstrated simultaneous reduction of the release of arachidonic acid metabolites, protein influx and leukocyte infiltration by cyclooxygenase, lipoxygenase inhibitors and steroids. Further studies to fully delineate the relationship between the clinical impact of anti-inflammatory compounds and their biochemical action would seem to be well advised.

Perspective

There is great diversity in the etiology of ocular inflammation, yet the ultimate clinical pathology shares similar characteristics. Each of the models of ocular inflammation provides an opportunity to explore special aspects of the biology of inflammation and also have utility in the evaluation of new anti-inflammatory compounds. The majority of attention would seem to have focussed upon the role of lipid mediators, derived from arachidonic acid, in models of nonimmunogenic and noninfectious ocular inflammation. In more recent years, there is a developing interest in the interplay between these lipid mediators and the immunological events of an inflammatory response. The available models of immunogenic ocular inflammation would appear to be well suited to such investigations.

The goal of research into ocular inflammation in animal models is very clear. Only by understanding the precise biochemical pathways and pharmacological actions of inflammatory mediators will we be able to expose new targets for drug intervention and develop more specific anti-inflammatory agents to ultimately replace the well-established corticosteroids.

References

1 Magendie F: De l'influence de la cinquième paire de nerfs sur la nutrition et les fonctions de l'œil. J Physiol Exp Pathol 1924;4:176–181.
2 Bernard C: Leçons sur la Physiologie et la Pathologie du Système Nerveux. 1858;2: 96–205.
3 Perkins ES: Influence of the fifth cranial nerve on the intraocular pressure of the rabbit eye. Br J Ophthalmol 1957;41:257–300.
4 Jampol LM, Axelrod A, Tessler H: Pathways of the eye's response to topical nitrogen mustard. Invest Ophthalmol 1976;15:486–489.
5 Butler JM, Unger WG, Hammond BR: Sensory mediation of the ocular response to neutral formaldehyde. Exp Eye Res 1979;28:577–589.
6 Camras CB, Bito LZ: The pathophysiological effects of nitrogen mustard on the rabbit eye. I. The biphasic intraocular pressure response and the role of prostaglandins. Exp Eye Res 1980;30:41–52.
7 Jampol LM, Noth J: Further studies of the ipsilateral and contralateral responses to topical nitrogen mustard. Exp Eye Res 1979;28:591–600.
8 Von Euler US, Gaddum JH: An unidentified depressor substance in tissue extracts. J Physiol (Lond) 1931;72:74–87.
9 Lembeck F: Zur Frage der zentralen Überfragung afferenter Impulse. III. Naunyn Schmiedebergs Arch Pharmacol 1953;219:197–213.
10 Pernow B: Substance P. Pharmacol Rev 1983;35:85–141.
11 Chang MM, Leeman SE, Niall HD: Amino acid sequence of substance P. Nature New Biol (Lond) 1971;232:86–87.
12 Tregear GW, Niall HD, Potts J, et al: Synthesis of substance P. Nature New Biol (Lond) 1971;232:87–89.

13 Otsuka M, Konishi S: Substance P – the first peptide neurotransmitter? Trends Neurosci 1983;6:317–320.
14 Olgart L, Gazelius B, Brodin E: Release of substance P-like immunoreactivity from dental pulp. Acta Physiol Scand 1977;80:510–512.
15 Bill A, Stjernschantz J, Mandahl A, et al: Substance P: Release on trigeminal nerve stimulation, effects in the eye. Acta Physiol Scand 1979;106:371–373.
16 Stjernschantz J, Sears M, Mishima H: Role of substance P in the antidromic vasodilation, neurogenic plasma extravasation and disruption of the blood-aqueous barrier in the rabbit eye. Naunyn Schmiedebergs Arch Pharmacol 1982;321:329–335.
17 Butler JM, Hammond BR: The effects of sensory denervation on the responses of the rabbit eye to prostaglandin E_1, bradykinin and substance P. Br J Pharmacol 1980;69:495–502.
18 Tervo K, Tervo T, Eranko L, et al: Immunoreactivity for substance P in the Gasserian ganglion, ophthalmic nerve and anterior segment of the rabbit eye. Histochem J 1981;13:435–443.
19 Tervo K, Tervo T, Eranko L, et al: Substance P – immunoreactive nerves in the human cornea and iris. Invest Ophthalmol Vis Sci 1982;23:671–674.
20 Ehinger B, Sundler F, Tervo K, et al: Substance P fibres in the anterior segment of the rabbit eye. Acta Physiol Scand 1983;118:215–218.
21 Unger WG, Butler JM, Cole DF, et al: Substance P, vasoactive intestinal peptide and somatostatin levels in ocular tissues of normal and sensorily denervated rabbit eyes. Exp Eye Res 1981;32:797–801.
22 Mandahl A, Bill A: Ocular responses to antidromic trigeminal stimulation, intracameral prostaglandin E_1 and E_2, capsaicin and substance P. Acta Physiol Scand 1981; 112:331–338.
23 Nishiyama A, Masuda K, Mochizuki M: Ocular effects of substance P. Jap J Ophthalmol 1981;25:362–369.
24 Bito LZ, Baroody RA, Backman A: The mechanism of peptidergic miosis. I. The structural basis of miotic potency among biologically active peptides. Curr Eye Res 1982;1:559–565.
25 Stjernschantz J, Sears M, Stjernschantz L: Intraocular effects of substance P in the rabbit. Invest Ophthalmol Vis Sci 1981;20:53–60.
26 Bito LZ, Nichols RR, Baroody RA: A comparison of the miotic and inflammatory effects of biologically active polypeptides and prostaglandin E_2 on the rabbit eye. Exp Eye Res 1982;34:325–337.
27 Nishiyama A, Masuda K, Mochizuki M: Ocular effects of vasoactive intestinal polypeptide. Invest Ophthalmol Vis Sci 1981;21(suppl):64.
28 Unger WG, Terenghi G, Ghatei MA, et al: Calcitonin-gene-related polypeptide as a mediator of the neurogenic ocular injury response. J Ocul Pharmacol 1985;1:189–199.
29 Malaty R, Close K, Franklin R, et al: The role of CGRP in intraocular inflammation. Invest Ophthalmol Vis Sci 1988;29(suppl):375.
30 Aronson SB, Goodner EK, Yamamoto BA, et al: Mechanisms of the host response of the eye. I. Changes in the anterior eye following immunization to a heterologous antigen. Arch Ophthalmol 1965;73:402–412.
31 Hall JM, Pribnow JF: Topical ocular immunization of rabbits. Invest Ophthalmol Vis Sci 1981;21:753–756.

32 Dwyer R, Turk JL, Darougar S: Immediate hypersensitivity in the guinea pig conjunctiva. I. Characterisation of the IgE and IgG antibodies involved. Int Arch Allergy Appl Immunol 1974;46:910–924.
33 Williams RN, Paterson CA, Eakins KE, et al: Quantification of ocular inflammation: Evaluation of polymorphonuclear leukocyte infiltration by measuring myeloperoxidase activity. Curr Eye Res 1983;2:465–470.
34 Friedlaender MH, Cyr RJ: Contact sensitivity in the guinea pig eye. Curr Eye Res 1981;1:403–407.
35 Segawa K, Smelser GK: Electron microscopy of experimental uveitis. Invest Ophthalmol 1969;8:497–520.
36 Allansmith MR, Baird RS, Block KJ: Degranulation of ocular mast cells in rats undergoing systemic anaphylaxis. Invest Ophthalmol Vis Sci 1980;19:1521–1524.
37 Walsh PW, Yunginger JW, Toni DG, et al: Topical ocular administration of chromolyn sodium for treatment in seasonal ragweed conjunctivitis. J Allergy Clin Immunol 1979;64:209–215.
38 Eakins KE, Bhattacherjee P: Histamine, prostaglandin and ocular inflammation. Exp Eye Res 1977;24:299–305.
39 Wirostko E, Bhattacherjee P, Eakins KE, et al: Cellular aspects of conjunctival inflammation induced by the synergistic action of histamine and prostaglandins. Immunopharmacology 1978;1:49–56.
40 Napoli S, Helm C, Feigen LP, et al: External ocular inflammatory effects of lipoxygenase enzyme products. Invest Ophthalmol Vis Sci 1986;27(suppl):322.
41 Spada CS, Woodward DF, Hawley SB, et al: Leukotrienes cause eosinophil emigration into conjunctival tissue. Prostaglandins 1986;31:795–809.
42 Woodward DF, Ledgard SE: Effect of LTD_4 on conjunctival vasopermeability and blood-aqueous barrier integrity. Invest Ophthalmol Vis Sci 1985;26:481–485.
43 Garceau D, Ford-Hutchinson AW: The role of leukotriene D_4 as a mediator of allergic conjunctivitis in the guinea pig. Eur J Pharmacol 1987;134:285–292.
44 Spada CS, Woodward DF, Hawley SB, et al: Synergistic effects of LTB_4 and LTD_4 on leukocyte emigration into the guinea pig conjunctiva. Am J Pathol 1988;130: 354–368.
45 Bisgaard H, Ford-Hutchinson AW, Charleston S: Production of peptido-leukotrienes in human tear fluid following antigen challenge. Prostaglandins 1984;28: 620–628.
46 Rahi AHS, Tripathi RC: Anatomy of passive Arthus reactions in the cornea. Mod Probl Ophthalmol 1976;16:165–172.
47 Elliot JH, Flax HM; Leibowitz HMI: The limbal cellular infiltrate in experimental corneal hypersensitivity. I. Morphological studies after primary sensitization. Arch Ophthalmol 1966;76:104–116.
48 Parks JJ, Leibowitz HMI, Harcmenee AE: A transient stage of delayed sensitivity during the early induction of immediate corneal sensitivity. J Exp Med 1962;115: 867–880.
49 Friedlaender MH, Dvorak HF: Morphology of delayed type hypersensitivity reactions in the guinea-pig cornea. J Immunol 1977;118:1558–1563.
50 Grabner G, Luger T, Smolin G, et al: Corneal epithelial cell-derived thymocyte-activating factor (CETAF). Invest Ophthalmol Vis Sci 1982;23:757–763.
51 Grabner G, Luger TA, Lugar BM: Biologic properties of the thymocyte-activating

factor (CETAF) produced by a rabbit corneal cell line (SIRC). Invest Ophthalmol Vis Sci 1983;24:589–595.
52 Leibowitz HM, Lass JH, Kupferman A: Quantitation of inflammation in the cornea. Arch Ophthalmol 1974;92:427–430.
53 Williams RN, Alvarez H, Paterson CA: Relative efficacy of anti-inflammatory corticosteroids in cornea. Cornea 1983;2:303–307.
54 Leibowitz HM, Kupferman A: Anti-inflammatory effectiveness in the cornea of topically administered prednisolone. Invest Ophthalmol 1974;13:757–763.
55 Bazan HEP: Effects of lipoxygenase inhibitors on the arachidonic acid cascade in the rabbit cornea after injury. J Ocul Pharmacol 1988;4:43–49.
56 Srinivasan BD: Corneal reepithelialization and anti-inflammatory agents. Trans Am Ophthalmol Soc 1982;80:758–822.
57 Kessler E, Mondino BJ, Brown SE: The corneal response to *Pseudomonas aeruginosa:* Histopathological and enzymatic characterization. Invest Ophthalmol Vis Sci 1977;16:116–125.
58 Kawaharajo K, Abe C, Homma JY, et al: Corneal ulcers caused by protease and elastase from *Pseudomonas aeruginosa.* Jap J Exp Med 1974;44:435–442.
59 Kreger AS, Gray LD: Purification of *Pseudomonas aeruginosa* proteases and microscopic characterization of pseudomonal protease-induced rabbit corneal damage. Infect Immun 1978;19:630–648.
60 Levinson RA, Paterson CA, Pfister RR: Ascorbic acid prevents corneal ulceration and perforation following experimental alkali burns. Invest Ophthalmol 1977;15: 986–993.
61 Dohlman CH, Pfister RR: Management of chemical burns of the eye. Transactions of New Orleans Academy of Ophthalmology, Symposium on the Cornea. St. Louis, Mosby, 1972, pp 105–109.
62 Pfister RR, McCulley JP, Friend J, et al: Collagenase activity of intact corneal epithelium in peripheral alkali burns. Arch Ophthalmol 1971;86:308–313.
63 Schwartzman ML, Abraham NG, Masferrer J, et al: Cytochrome P-450 dependent arachidonate metabolism in corneal epithelium: formation of two biologically active compounds; in Samuelsson B, Ramwell P, Padetti R (eds): Advances in Prostaglandin, Thromboxane and Leukotriene Research. New York, Raven Press, 1987, pp 78–83.
64 Masferrer RL, Schwartzman ML: Novel cytochrome P-450 dependent arachidonic acid metabolites and their ocular effects; in Bito LZ, Stjernschantz J (eds): The Ocular Effects of Prostaglandins and Other Eicosanoids. New York, Liss, 1989, vol 36, pp 85–93.
65 Bazan HEP, Reddy STK, Woodland JM, et al: The accumulation of platelet-activating factor in the injured cornea may be interrelated with the synthesis of lipoxygenase products. Biochem Biophys Res Commun 1987;147:915–920.
66 Braquet P, Shen TY, Touqui L, et al: Perspectives in platelet-activating factor research. Pharmacol Rev. 1987;39:97–145.
67 Eakins KE, Whitelock RAF, Perkins ES, et al: Release of prostaglandin in ocular inflammation in the rabbit. Nature 1972;239:248–249.
68 Baldwin HA, Borgmann AR: An improved immunogenic uveitis test in rabbits for evaluation of anti-inflammatory compounds. Proc Soc Exp Biol Med 1970;133: 1326–1330.

69 Kulkarni PS, Bhattacherjee P, Eakins KE, et al: Anti-inflammatory activity of betamethasone phosphate, dexamethasone phosphate and indomethacin in rabbit ocular inflammation induced by bovine serum albumin. Curr Eye Res 1981;1:43–47.

70 Collins RC: Experimental studies on sympathetic ophthalmia. Am J Ophthalmol 1949;32:1687–1699.

71 Aronson SB, Hogan MJ, Zweigart P: Homoimmune uveitis in the guinea pig. I. General concepts of auto- and homoimmunity, methods and manifestations. Arch Ophthalmol 1963;69:105–109.

72 Aronson SB: Hypersensitivity reactions in the uveal tract caused by uveal antigen-antibody reactions; in Maumenee AE, Sivlerstein AM (eds): Immunopathology of Uveitis. Baltimore, Williams & Wilkins, 1964, pp 182–191.

73 Aronson SB, Hogan MJ, Zweigart P: Homoimmune uveitis in guinea pigs. II. Clinical manifestations. Arch Ophthalmol 1963;69:203–207.

74 Wacker WB, Lipton MM, Ongchua FE: Antibody production in the guinea pig to homologous uvea. Proc Soc Exp Biol Med 1964;117:150–154.

75 Wacker WB, Lipton MM: Experimental allergic uveitis. Homologous retina as uveitogenic antigen. Nature 1965;206:2530–2534.

76 Wacker WB, Donoso LA, Kalsow CM, et al: Experimental allergic uveitis. Isolation, characterization and localization of a soluble antigen from bovine retina. J. Immunol 1977;119:1949–1958.

77 Wacker WB: Retinal autoimmunity and pathogenesis of uveitis; in Helmsen R, Suran AA, Gery I, et al (eds): Immunology of the Eye, Workshop II. Washington, IRL Press, 1981, pp 11–31.

78 Faure J-P, de Kozak Y: Cellular and humoral reactions to retinal antigen: Specific suppression of experimental uveo-retinitis; in Helmsen R, Suran AA, Gery I, et al (eds): Immunology of the Eye, Workshop II. Washington, IRL Press, 1981, pp 33–48.

79 Nussenblatt RB, Gery I, Kuwabara T, et al: The role of retinal S-antigen in primate uveitis; in Helmsen R, Suran AA, Gery I, et al (eds): Immunology of the Eye, Workshop II. Washington, IRL Press, 1981, pp 49–65.

80 Eakins KE: Prostaglandins and the eye; in Karim SM (ed): Prostaglandins: Physiological, Pharmacological and Pathological Aspects. Baltimore, University Park Press, 1976, pp 63–81.

81 Parker JA, Goetzl ED, Friedlaender MH: Leukotrienes in aqueous humor of patients with uveitis. Arch Ophthalmol 1986;104:722–724.

82 Pettipher ER, Higgs G, Henderson B: Interleukin 1 induced leukocyte infiltration and cartilage proteoglycan degradation in the synovial joint. Proc Natl Acad Sci USA 1986;83:8749–8753.

83 Dayer JM: Mononuclear cell factor – interleukin-1 in rheumatoid arthritis. Br J Rheumatol 1985;24(suppl 1):15–20.

84 Bernheim HA, Gilbert TM, Stitt TT: Prostaglandin E levels in third ventricular cerebrospinal fluid of rabbits during fever and changes in body temperature. J Physiol (Lond) 1980;301:69–78.

85 Dinarello CA: Interleukin 1. Rev Infect Dis 1984;6:51–94.

86 Prendergast RA, Libre PE, Orencole SF, et al: Ocular IL-1 generation in immunogenic uveitis. Invest Ophthalmol Vis Sci 1988;29(suppl):65.

87 Bhattacherjee P, Henderson B: Inflammatory responses to intraocularly injected interleukin 1. Curr Eye Res 1987;6:929–934.

88 Rosenbaum RT, Samples JR, Howes EL: Neovascularization and inflammation induced by tumor necrosis factor in the rabbit eye. Invest Ophthalmol Vis Sci 1988; 29(suppl):99.
89 Howes EL, Aronson SB, McKay DG: Ocular vascular permeability. Effect of systemic administration of bacterial endotoxin. Arch Ophthalmol 1970;84:360–367.
90 Bhattacherjee P: Release of prostaglandin-like substances by Shigella endotoxin and its inhibition by non-steroidal anti-inflammatory compounds. Br J Pharmacol 1975; 54:489–494.
91 Hammond BR, Bhattacherjee P: A method for the intraocular injection of microvolumes in small animals. Curr Eye Res 1981;1:683–685.
92 Bito LZ: The effect of experimental uveitis on anterior or uveal prostaglandin transport and aqueous humour composition. Invest Ophthalmol 1974;13:959–966.
93 Fleisher LN, McGahan MC: Time course for prostaglandin synthesis by rabbit lens during endotoxin-induced ocular inflammation. Curr Eye Res 1986;5:629–634.
94 Csukas S, Paterson CA, Brown K, et al: Temporal relationship between arachidonic acid metabolite release and physiological response during endotoxin-induced inflammation in the rabbit eye. Invest Ophthalmol Vis Sci, submitted.
95 Rosenbaum JT, McDevitt HO, Guss RB, et al: Endotoxin-induced uveitis in rats as a model for human disease. Nature (Lond) 1980;286:611–613.
96 Bhattacherjee P, Williams RN, Eakins KE: An evaluation of ocular inflammation following the injection of bacterial endotoxin into the rat footpad. Invest Ophthalmol Vis Sci 1983;24:196–202.
97 Howes EL, Goldyne ME, Perez HD, et al: Lipopolysaccharide tolerance inhibits eye inflammation. I. Reduced immune complex or lipopolysaccharide effects. Arch Ophthalmol 1985;103:257–260.
98 Herbort CP, Okumura A, Mochizuki M: Immunopharmacological analysis of endotoxin-induced uveitis in rats. Exp Eye Res, in press.
99 Kulkarni PS, Bhattacherjee P: Effects of selective cyclooxygenase and lipoxygenase inhibitors on the ocular injury and inflammatory responses. Invest Ophthalmol Vis Sci 1989;30, in press.
100 Williams RN, Paterson CA: The influence of topical corticosteroid therapy upon polymorphonuclear leukocyte distribution, vascular integrity and ascorbate levels in endotoxin-induced inflammation of the rabbit eye. Exp Eye Res 1987;44:191–198.
101 Bhattacherjee P, Williams RN, Eakins KE: A comparison of the ocular anti-inflammatory activity of steroidal and non-steroidal compounds in the rat. Invest Ophthalmol Vis Sci 1983;24:1143–1146.
102 Rosenbaum JT, Raymond W, Seymour BWP, et al: The effect of corticosteroids or nitrogen mustard on aqueous humor chemotactic activity induced by intravitreal endotoxin. Clin Immunol Immunopathol 1986;39:414–420.
103 Bhattacherjee P: The role of arachidonate metabolites in ocular inflammation; in Bito LZ, Stjernschantz J (eds): The Ocular Effect of Prostaglandins and Other Eicosanoids. New York, Liss, 1989, pp 211–227.

P. Bhattacherjee, PhD, Department of Ophthalmology and Visual Sciences,
University of Louisville, Kentucky Lions Eye Research Institute,
301 E. Muhammad Ali Blvd., Louisville, KY 40202–1511 (USA)

Bazan NG (ed): Lipid Mediators in Eye Inflammation.
New Trends Lipid Mediators Res. Basel, Karger, 1990, vol 5, pp 83–93

The Role of Phospholipase A_2 in Experimental Autoimmune Uveoretinitis

Kaj Mahlberg[a], *Hannu Uusitalo*[b], *Arto Palkama*[b], *Thomas Tallberg*[a]

[a] Laboratory of Immunology, Helsinki University Central Hospital, and
[b] Department of Anatomy, Eye Research Laboratory, University of Helsinki, Finland

Retinal S-Antigen

It is known since the early works of Wacker and Lipton [1] that retina is a rich source of immunopathogenic autoantigens capable of eliciting an autoimmune inflammation in the retina and uvea. Although several autoantigens have been found in the retina, the first and best known of these autoantigens is the retinal soluble antigen, called S-antigen. This was first isolated independently by two separate laboratories in the mid-70s [2–4]. This autoantigen is capable of inducing an experimental autoimmune uveoretinitis (EAU) [4].

S-antigen was found to be a 48-kd protein regulating light-dependent phosphodiesterase [5] and localized in the rod photoreceptor cells [4, 6, 7] (fig. 1). The amino acid sequence of bovine and human S-antigen has been determined by cDNA sequencing [8]. Interestingly, S-antigen seems to share sequence homology with alpha-transducin including the putative rhodopsin- and phophoryl-binding sites [9, 10]. Although there is a strong immunological cross-reactivity between retinal S-antigens isolated from different species, they are not identical to each other [10, 11].

The epitopes of S-antigen have been characterized by Gregerson et al. [12, 13] using cyanogen bromide peptides. They demonstrated that CB 123

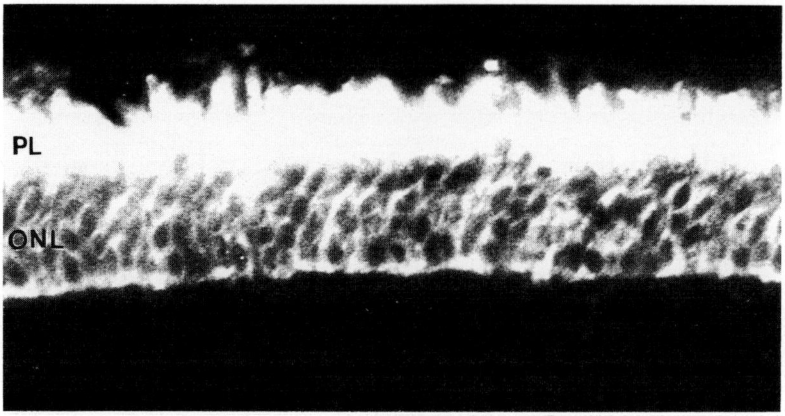

Fig. 1. Immunohistochemical demonstration of retinal S-antigen in guinea pig retina. Bright fluorescence has been seen in the photoreceptor cell being most intense in the outer segments. PL = Photoreceptor layer; ONL = outer nuclear layer.

peptide was both antigenic and uveitogenic and produced a uveitogenic, CB 123-specific cell line, establishing the cell-mediated nature of EAU. Donoso et al. [14, 15] have studied the immunopathogenic epitopes of S-antigen using synthetic polypeptides. They found first an M peptide (an 18-amino acid polypeptide of bovine S-antigen) and later two synthetic polypeptides of human S-antigen (HSA 319 and HSA 320) to be capable of inducing EAU.

Retinal S-antigen has been thought to have an etiological role in clinical cases of uveitis [16–18], e.g. sympathetic ophthalmia and birdshot retinochoroidopathy [19]. Circulating antibodies to S-antigen has been demonstrated in different forms of uveitis [17, 18, 20]. In contrast to these observations, Doekes et al. [21] did not find any change in antibody titers directed against human S-antigen in patients suffering from uveitis. Cell-mediated immunity and T cells seem to play a dominant role in the induction of EAU [12, 22]. Cellular immune responsiveness to S-antigen has been demonstrated in patients suffering from uveitis [21, 23]. Furthermore, EAU can be mediated by activated T cell lines to virgin animals [12, 22]. However, B cells and antibodies to S-antigen seem to have a role in the regulation of this type of autoimmune disease [24].

Experimental Autoimmune Uveoretinitis

Active immunization with retinal extract and purified S-antigen was found to induce inflammation in the eye in several species of experimental animals [1, 4]. EAU is clinically characterized by iris and pericorneal hyperemia, followed by inflammatory exudates in the anterior chamber and vitreous body. Histopathological changes included destruction of photoreceptor cells in the retina, focal infiltration of inflammatory cells in the choroid as well as infiltration of mono- and polymorphonuclear cells in the ciliary body and iris [4, 23, 25]. However, the clinical and histopathological characteristics seem to be very dependent on the species and dose of S-antigen used [23, 26–28]. The susceptibility of different species and strains to EAU seems to be correlated with the presence and number of choroidal mast cells [29]. The inflammation in the guinea pigs induced with a relatively low dose (10–25 μg) of S-antigen induces a relatively mild inflammation in the retina, with focal infiltration of inflammatory cells in the choroid and a moderate inflammation in the anterior segment of the eye [25, 26]. This inflammation resembles the clinical as well as histopathological characteristics of sympathetic ophthalmia [26]. This particular type of inflammation has been used as a model for experimental uveitis in our series of experiments. The intensity of the inflammation has been characterized both clinically using a slitlamp, and histopathologically using conventionally stained paraffin serial sections [25].

The Role of Phospholipase A_2 in the Arachidonic Acid Cascade

Arachidonic acid is usually the substrate for both the cyclooxygenase and the lipoxygenase pathway [30, 31]. However, the concentration of arachidonic acid is in most cells very low and the limiting step in the synthesis of both cyclooxygenase and lipoxygenase products is the availability of free arachidonate [32, 33]. Agents that activate the arachidonic acid cascade seem to act by releasing arachidonate from membrane phospholipids. One mechanism is the activation of phospholipase A_2 (PLA_2), which by no means is the only enzyme hydrolyzing phospholipids into free arachidonic acid [34]. Various stimuli are capable of activating PLA_2 [35–38]. This stimulation is different for each cell type. Some of them work through receptor-mediated mechanisms [39, 40]. Other mechanisms, such as the ionophore A23187, cause a nonreceptor-mediated stimulation of arachidonic acid release [41]. The mechanism of release of arachidonate is probably the same in all cell types [42]. Activation of specific phospholipases is the first step in this release. By definition, PLA_2 encompasses enzymes which hydrolyze the ester linkage in position 2 of phosphoglycerides. They can be divided into three different classes according to the source in which they are found: (1) venoms; (2) digestive exocrine glands, or (3) as intracellular enzyme occurring in tissues or microorganisms. Intracellular PLA_2 has been isolated from mitochondria, microsomes and lysosomes. A common feature of most PLA_2s is the requirement of calcium and its resistance to heat [43].

The activation PLA_2 is suggested to be initiated by methylation of phosphatidylethanolamine to form phosphatidylcholine [44]. This methylation is coupled to calcium influx into cells which in turn activates PLA_2. Phosphatidylcholine serves in turn as a substrate for the calcium-activated PLA_2 [42, 44]. PLA_2 seems to be present in most inflammatory disorders as well as in anaphylactic reactions [45]. Activation of PLA_2, for example in the pancreas, is probably one of the major factors leading to hemorrhagic pancreatitis [46, 47]. In anaphylaxis the complexing of the antibody with antigen on the cell surface renders the plasma membrane permeable to extracellular calcium ions resulting in a immediate rise of free intracellular Ca^{++}. This is essential for the triggering of the arachidonic acid cascade. The liberation of free arachidonic acid from membrane phospholipids is also in this case catalyzed, among others, by PLA_2 [48].

In the eye, both the anterior uvea and the conjunctiva possess a marked capacity for prostaglandin synthesis [49, 50]. For example, mechanical irritation of the iris results in the release of prostaglandins [51, 52]. Also, intracameral injection or topical application of prostaglandins to the rabbit eye causes ocular inflammatory responses in the anterior uvea including increased permeability in the uveal blood vessels [53, 54]. Furthermore, an experimental uveitis caused by intravitreal bovine serum albumin has been shown to induce prostaglandin production in the eye [55].

The role of leukotrienes in ocular tissue is not clearly understood although the iris and the conjunctiva of many species are able to synthesize compounds of the lipoxygenase pathway [56, 57]. In the eye, the injection of LTB_4 causes accumulation of polymorphonuclear leukocytes in the aqueous humor [58, 59], while the effects of LTC_4 and LTD_4 are minor. These compounds are reported to cause a long-lasting miosis and a moderate reduction in the blood flow of the anterior uvea without damaging the blood-aqueous barrier [60].

PLA_2 in EAU

When 10–25 µg of freshly prepared bovine S-antigen mixed with Freund's complete adjuvant was injected subcutaneously into test animals, most guinea pigs developed active uveitis within 14 days and a significant increase in PLA_2 levels in aqueous humor could be recorded (table 1). This activity seemed furthermore to be dependent upon the severity of the ocular inflammation since there was a clear correlation between these two parameters. Surprisingly, eyes with a mild form of EAU showed significant elevation of PGE_2 and LTC_4 while in eyes with severe uveitis these mediators could not be detected [28].

The exact biochemical mechanism leading to arachidonate mobilization is not completely understood, but these results could suggest that in experimental EAU, initiation partly depends on PLA_2 activation with a possible early increase in PGE_2 and LTC_4. The rapid inactivation of these compounds could furthermore suggest that these mediators probably could act locally having inflammatory and/or anti-inflammatory effects. The role of PLA_2 is to liberate arachidonic acid from the cell membrane for the cyclooxygenase and lipoxygenase pathways which in part leads to destruction of cell membranes in ocular tissues.

PLA_2 activity correlated with the clinical grading of EAU [28]. When PLA_2 levels were correlated with the histopathological findings in ocular tissues, it was noticed that the best correlation was obtained with mono- and polymorphonuclear cell infiltration in the iris. No correlation was

Table 1. Biochemical parameters measured from aqueous humor and iris-ciliary body

	EAU			Control			Level of signif-
	mean	SE	n	mean	SE	n	icance
PLA_2, µg fatty acid/50 µl/18 h	23.25	5.15	19	4.68	1.84	8	$p < 0.01$
LTC_4, ng/0.1 ml	0.29	0.13	50	0.08	0.04	27	n.s.
PGE_2, ng/0.5 ml	0.54	0.17	12	0.00	0.00	4	$p < 0.01$
MPO, units/mg iris-ciliary homogenate	0.23	0.54	64	0.0003	0.00002	25	$p < 0.001$

PLA_2 and PGE_2 measured from aqueous humor and myeloperoxidase (MPO) measured from iris-ciliary body homogenate were significantly elevated in the trial groups as compared to the controls.

found between LTC_4 or PGE_2 levels in the aqueous humor and the histopathological grading [25]. Whether PLA_2, LTC_4 and PGE_2 originated in the anterior segment of the eye, or in inflammatory cells present in these tissues and aqueous humor, is not clear.

Inhibition of EAU and PLA_2

Since PLA_2 was elevated in aqueous humor of eyes during EAU, it seemed natural to try to prevent PLA_2 activation. Corticosteroids with their known side effects are the principal remedies for the treatment of ocular inflammations today. Corticosteroids are known to induce proteins called lipocortins [61–63] which are proteins with phospholipase inhibitory properties [64]. Since PLA_2 activation is dependent on calcium, disodium EDTA, a chelating agent which forms complexes with divalent and trivalent cations, was applied as an alternative to corticosteroids. The effect of disodium EDTA was furthermore compared with that of ciclosporin A, a selective T-cell inhibitor (Interleukin-2) [65] used in the treatment of EAU in many studies [66, 67].

Table 2. Suppression of EAU

	EAU control	Methylprednisolone (s.c.)	Dexamethasone (topical)	Ciclosporin (s.c.)	EDTA (topical)	EDTA (s.c.)
Clinical grading	4.85	2.91	2.64	2.35	2.15	1.75
SD	1.9	1.6	2.5	1.5	1.7	1.2
n	26	22	14	20	20	20
PLA_2, µg fatty acid/18 h	19.2	9.7	13.7	4.61	9.1	5.3
SD	31.1	11.8	15.6	1.7	10.5	4.2
n	24	22	14	19	20	20
MPO, units/mg	0.51	0.69	1.12	0.08	0.55	0.09
SD	0.8	1.5	1.4	0.2	1.2	0.1
n	25	22	14	19	20	16

Clinically the induced disease was significantly decreased in all treatment groups. The best suppression was obtained in the group treated subcutaneously with disodium EDTA. PLA_2 levels in aqueous humor as well as myeloperoxidase (MPO) activity measured from iris-ciliary body homogenate was significantly decreased in the groups treated with ciclosporin A and disodium EDTA.

The best suppression of the inflammatory reaction was obtained in experimental animals treated subcutaneously with disodium EDTA. Treatment with corticosteroids also caused a decrease in the clinical grading but was not as effective as disodium EDTA (table 2). Only subcutaneous disodium EDTA and ciclosporin A significantly prevented the increase in PLA_2 activity as measured in aqueous humor. Both drugs also decreased the humoral response to S-antigen [68].

Although the mechanism of action of disodium EDTA is not yet clear, one can speculate that it might prevent the rise in the intracellular free Ca^{2+} concentration [45, 69] seen during the activation of PLA_2. Thus, using compounds that chelate extracellular divalent cations, the activation of the arachidonic acid cascade could probably be prevented. Interestingly, ciclosporin A also suppresses, in a dose-dependent manner, PLA_2 activity and the release of PGE_2 in normal human lymphocytes [70].

PLA_2 seems to be related with the induction of EAU. Furthermore, efficient treatments suppressing the inflammatory reactions caused by S-antigen are able to reduce PLA_2 activity in the eye. Disodium EDTA has been proven to be effective in reducing both EAU and PLA_2 activity. Whether disodium EDTA will have a role in the treatment of clinical forms of uveitis remains to be seen.

References

1 Wacker WB, Lipton MM: Experimental allergic uveitis. Homologous retina as uveitogenic antigen. Nature 1965;206:253–257.

2 Wacker WB, Donoso LA: Characterization of the uveitopathic antigen-induced isolate from the retina. ARVO Abstracts. Invest Ophthalmol Vis Sci 1975(suppl):94.

3 Faure JP, de Kozak Y, Dorey C, et al: Experimental auto-immune uveo-retinitis. Activity of different antigenic preparations from the retina. ARVO Abstracts. Invest Ophthalmol Vis Sci 1975(suppl):94.

4 Wacker WB, Donoso LA, Kalsow CM, et al: Experimental autoimmune uveitis. Isolation, characterization and localization of a soluble uveitopathogenic antigen from bovine retina. J Immunol 1977;119:1949–1958.

5 Pfister C, Chabre M, Plouet J, et al: Retinal S-antigen identified as the 48 K protein regulating light-dependent phosphodiesterase in rods. Science 1985;228:891–893.

6 Yajima S, Seki F, Takano S, et al: Localization of S-antigen by enzyme-labeled antibody method and electron microscopy. Jap J Ophthalmol 1983;27:526–534.

7 Uusitalo H, Lehtosalo JI, Gregerson DS, et al: Ultrastructural localization of retinal S-antigen in the rat. Arch Clin Exp Ophthalmol 1985;222:118–122.

8 Shinohara T, Dietzschold B, Craft CM, et al: Primary and secondary structure of

bovine retinal S-antigen (48 K protein). Proc Natl Acad Sci USA 1987;84:6975–6979.
9 Wistow GJ, Katial A, Craft C, et al: Sequence analysis of bovine retinal S-antigen. Relationships with alpha-transducin and G-proteins. FEBS Lett 1986;196:23–28.
10 Yamaki K, Tsuda M, Shinohara T: The sequence of human retinal S-antigen reveals similarities with alpha-transducin. FEBS Lett 1988;234:39–43.
11 Wacker WB: Isolation of retinal S-antigen and its role in experimental ocular autoimmunity; in Sears ML (ed): New Directions in Ophthalmic Research. London, Yale University Press, 1981, pp 5–29.
12 Gregerson DS, Obritsch WF, Fling SP, et al: S-antigen specific rat T cell lines recognize peptide fragments of S-antigen and mediate experimental autoimmune uveoretinitis and pinealitis. J Immunol 1986;136:2875–2882.
13 Gregerson DS, Obritsch WF, Fling ST: Identification of a uveitogenic cyanogen bromide peptide of bovine retinal S-antigen and preparation of a uveitogenic, peptide-specific T cell line. Eur J Immunol 1987;17:405–411.
14 Donoso LA, Merryman CF, Sery TW, et al: S-antigen: characterization of a pathogenic epitope which mediates experimental autoimmune uveitis and pinealitis in Lewis rats. Curr Eye Res 1987;6:1151–1159.
15 Donoso LA, Yamaki K, Merryman CF, et al: Human S-antigen: characterization of uveitopathogenic sites. Curr Eye Res 1988;7:1077–1085.
16 Nussenblatt RB, Gery I, Ballintine EJ, et al: Cellular immune responsiveness of uveitis patients to retinal S antigen. Am J Ophthalmol 1980;89:173–179.
17 Gregerson DS, Abrahams IW, Thirkill CE: Serum antibody levels of uveitis patients to bovine retinal antigens. Invest Ophthalmol Vis Sci 1981;21:669–680.
18 Uusitalo RJ, Stjernschantz J, Mahlberg K, et al: Serum antibody level to S-antigen in children with chronic uveitis. Br J Ophthalmol 1985;69:212–216.
19 Nussenblatt RB, Mittal KK, Ryan S, et al: Birdshot retinochoroidopathy associated with LHA-A29 antigen and immune responsiveness to retinal S-antigen. Am J Ophthalmol 1982;94:147–158.
20 Faure J-P, Kozak Y: Cellular and humoral reactions to retinal antigen, specific suppression of experimental uveoretinitis; in Helmsen RJ, Suran A, Nussenblatt RB (eds): Immunology of the Eye: Workshop II. London, IRL Press, 1981, pp 33–48.
21 Doekes G, van der Gaag R, Rothova A, et al: Humoral and cellular immune responsiveness to human S-antigen in uveitis. Curr Eye Res 1987;6:909–919.
22 Caspi RR, Roberge FG, McAllister CG, et al: T cell lines mediating experimental autoimmune uveoretinitis (EAU) in the rat. J Immunol 1986;136:928–933.
23 Nussenblatt RB, Kuwabara T, de Monasterio FM, et al: S-antigen uveitis in primates. A new model for human disease. Arch Ophthalmol 1981;99:1090–1092.
24 de Kozak Y, Mirshahi M, Boucheix C, et al: Prevention of experimental autoimmune uveoretinitis by active immunization with autoantigen specific monoclonal antibodies. Eur J Immunol 1987;17:541–547.
25 Mahlberg K, Uusitalo H, Palkama A, et al: Correlation between histopathological, clinical and biochemical parameters in S-antigen induced esperimental autoimmune uveitis in guinea pigs. Exp Eye Res 1987;45:157–167.
26 Rao NA, Wacker WB, Marak GE Jr: Experimental allergic uveitis: Clinicopathologic features associated with varying doses of S-antigen. Arch Ophthalmol 1979;97:1954–1958.

27 Kalsow CM, Wacker WB: Rabbit ocular and pineal autoimmune response to retinal antigens. Curr Eye Res 1986;5:579–586.
28 Mahlberg K, Uusitalo R, Palkama A, et al: Phospholipase A_2, leukotriene C_4 and prostaglandin E_2 levels in aqueous humor of guinea pigs with experimental S-antigen induced autoimmune uveitis. Curr Eye Res 1987;6:321–335.
29 Mochizuki M, Kuwabara T, Chan CC, et al: An association between susceptibility to experimental autoimmune uveitis and choroidal mast cell numbers. J Immunol 1984;133:1699–1701.
30 Lands WEM, Samuelsson B: Phospholipid precursors of prostaglandins. Biochim Biophys Acta 1968;164:426–429.
31 Samuelson B: Leukotrienes: Mediators of immediate hypersensitivity reactions and inflammation. Science 1983;220:568–575.
32 Flower FJ, Blackwell GJ: The importance of phospholipase-A_2 in prostaglandin biosynthesis. Biochem Pharmacol 1976;25:285–291.
33 Kunze H, Vogt W: Significance of phospholipase A for prostaglandin formation. Ann NY Acad Sci 1984;180:123–125.
34 Irvine RF: How is the level of free arachidonic acid controlled in mammalian cells? Biochem J 1982;204:3–16.
35 McGiff JC, Terrango NA, Malik KU, et al: Release of a prostaglandin E-like substance from canine kidney by bradykinin. Comparison with eledoisin. Circ Res 1972;31:36–43.
36 Vargaftig B, Dao Hai N: Selective inhibition by mepacrine of the release of 'rabbit aorta contraction substance' evoked by the administration of bradykinin. J Pharm Pharmacol 1972;24:159–161.
37 Rillema J, Wild E: Prolactin activation of phospholipase A activity in membrane preparations from mammary glands. Endocrinology 1977;100:1219–1222.
38 Chang J, Gillman SC, Lewis AJ: Interleukin 1 activates phospholipase A_2 in rabbit chondrocytes: a possible signal for IL-1 action. J Immunol 1986;136:1283–1287.
39 Askenase PW, Bursztan S, Gershon MD, et al: T cell dependent mast cell degranulation and release of serotonin in murine delayed-type hypersensitivity. J Exp Med 1980;152:1358–1374.
40 Lewis RA, Holgata ST, Roberts LJ, et al: Preferred generation of prostaglandin D_2 by rat and human mast cells; in Becker EL, Simon AS, Auster KF, et al (eds): Biochemistry of the Acute Allergic Reactions. Proceedings of the 4th International Symposium. New York, Liss, 1982.
41 Rittenhouse-Simmons S, Deykin D: The mobilization of arachidonic acid in platelets exposed to thrombin or ionophore A23187. J Clin Invest 1977;60:495–498.
42 Stenson WF, Parker CW: Leukotrienes; in Stollerman HG, Harrington WJ, Lamont JT, et al (eds): Advances in Internal Medicine. Chicago, Yearbook Medical Publishers, 1984, pp 175–192.
43 Brockerhoff H, Jensen RG: Lipolytic Enzymes. New York, Academic Press, 1974.
44 Hirata F, Axelrod J: Phospholipid methylation and biological signal transmission. Science 1980;209:1082–1084.
45 Bieven MA, Roberts J, Moore JP, et al: The mechanism of calcium signal and the correlation with histamine release. J Biol Chem 1984;259:7129–7136.
46 Tykkä H, Mahlberg K, Pantzar P, et al: Phospholipase A_2 inhibitors and their possible clinical use in treatment of acute pancreatitis. Scand J Gastroenterol 1980;15: 519–528.

47 Tykkä HT, Vaittinen EJ, Mahlberg K, et al: A randomized double-blind study using $CaNa_2$ EDTA, a phospholipase A_2 inhibitor, in the management of human acute pancreatitis. Scand J Gastroenterol 1985;20:5–12.

48 Zor U, Harel T, Ashkenazy Y, et al: The role of calcium in leukotriene production: Immune response induces cardiac myolysis; in Zor U, Naor Z, Kohen F (eds): Advances in Prostaglandin, Thromboxane and Leukotriene Research. New York, Raven Press, 1986, vol 16, pp 269–286.

49 Bhattacherjee P, Kulkarni PS, Eakins KE: Metabolism of arachidonic acid in rabbit ocular tissues. Invest Ophthalmol Vis Sci 1979;18:173–178.

50 Kulkarni PS, Srinivasan BD: Synthesis of slow reactive substance-like activity in rabbit conjunctiva and anterior uvea. Invest Ophthalmol Vis Sci 1983;24:1079–1085.

51 Ambache N, Kavanagh L, Whiting J: Effect of mechanical stimulation on rabbits' eyes: release of active substances in anterior chamber perfusates. J Physiol (Lond) 1965;176:378–408.

52 Cole DF, Unger WG: Prostaglandins as mediators for the responses of the eye to trauma. Exp Eye Res 1973;17:357–368.

53 Beitch RR, Eakins KE: The effect of prostaglandins on the intraocular pressure of the rabbit. Br J Pharmacol 1969;37:158–167.

54 Chiang TS, Thomas RP: Consensual ocular hypertensive response to prostaglandins. Invest Ophthalmol Vis Sci 1972;11:169–176.

55 Eakins KE, Whitelocke RAF, Perkins ES, et al: Release of a prostaglandin in ocular inflammation. Nature 1972;239:248–249.

56 Bhattacherjee P, Kulkarni PS, Eakins KE: Differential inflammatory effect of arachidonic acid on rabbit conjunctiva and iris: A possible role of lipoxygenase in the conjunctival response. Adv Prostaglandin Thromboxane Res 1980;8:1727–1731.

57 Williams RN, Bhattacherjee P, Eakins KE: Biosynthesis of lipoxygenase products of ocular tissues. Exp Eye Res 1983;36:397–402.

58 Bhattacherjee P, Hammond B, Salmon JA, et al: Chemotactic response to some arachidonic acid lipoxygenase products in the rabbit eye. Eur J Ophthalmol 1981; 73:21–28.

59 Stjernschantz J, Sherk T, Borgeat P, et al: Intraocular effects of lipoxygenase pathway products in arachidonic acid metabolism. Acta Ophthalmol 1984;62:104–111.

60 Stjernschantz J, Sherk T, Sears ML: Ocular responses to leukotriene C_4 and D_4 in the cat. Prostaglandins 1984;27:5–16.

61 Flower RJ, Blackwell GJ: Anti-inflammatory steroids induce biosynthesis of a phospholipase A_2 inhibitor which prevents prostaglandin generation. Nature 1978;278: 456–459.

62 Blackwell GJ, Carnuccio R, DiRosa M, et al: Macrocortin: a polypeptide causing the anti-phospholipase effect of glucocorticoids. Nature 1980;287:147–149.

63 Carnuccio R, DiRosa M, Flower RJ, et al: The inhibition by hydrocortisone of prostaglandin biosynthesis in rat peritoneal leukocytes is correlated with intracellular macrocortin levels. Br J Pharmacol 1981;74:322–324.

64 Flower RJ: Lipocortins; in Samuelsson B, Paoletti R, Ramwell P (eds): Advances in Prostaglandin, Thromboxane, and Leukotriene Research. New York, Raven Press, 1978, vol 17A, pp 577–580.

65 White DJ, Plumb AM, Pawelec G, et al: Cyclosporin A: an immunosuppressive agent preferentially active against T cells. Transplantation 1979;27:55–58.
66 Nussenblatt RB, Rodrigues MM, Wacker WB, et al: Cyclosporin A. Inhibition of experimental autoimmune uveitis in Levis rats. J Clin Invest 1981;67:1228–1231.
67 Nussenblatt RB, Rodrigues MM, Salinas-Carmona MC, et al: Modulation of experimental autoimmune uveitis with cylcosporin A. Arch Ophthalmol 1982;100:1146–1149.
68 Mahlberg K, Uusitalo H, Uusitalo R, et al: Suppression of experimental autoimmune uveitis in guinea pigs by ethylenediaminetetraacetic acid, corticosteroids, and cyclosporin. J Ocul Pharmacol 1987;3:199–210.
69 Ishizaka T: The Robert A. Cooke Memorial Lecture: Analysis of triggering event in mast cells for immunoglobulin E-mediated histamine release. J Allergy Clin Immunol 1981;67:90–96.
70 Niwa Y, Kano T, Taniguchi S, et al: Effect of cyclosporin A on the membrane-associated events in human leukocytes with special reference to the similarity with dexamethasone. Biochem Pharmacol 1986;35:947–951.

Kaj Mahlberg, Laboratory of Immunology, Helsinki University Central Hospital, Mannerheimintie 172, SF–00280 Helsinki (Finland)

Bazan NG (ed): Lipid Mediators in Eye Inflammation.
New Trends Lipid Mediators Res. Basel, Karger, 1990, vol 5, pp 94–109

Ocular Inflammation Induced by Lens Protein and Its Prevention with New Agents

George C.Y. Chiou, Min S. Chang

Department of Medical Pharmacology and Toxicology,
Institute of Ocular Pharmacology, Texas A & M University College of Medicine,
College Station, Tex., USA

Inflammation can occur in any vascularized living tissue of the body. These tissues exhibit all the cardinal signs described by Celsus: rubor, tumor, calor, and dolor (redness, swelling, heat and pain). The inflammatory process is a normal defense mechanism which serves to dilute, remove or wall-off injurious agents. At the same time, the inflammatory process is closely intertwined with the healing process. The repair of injured tissue begins in the early phases of inflammation and reaches completion after the injurious agent has been neutralized or removed.

The inflammatory process can be thought of as a two-edged sword. It cuts through injurious agents by removing them or isolating them and sets in motion the processes of regeneration and repair. On the other hand, the inflammatory process may inflict damages of its own on the host. This is where a fifth cardinal sign, loss of function, as observed by Virchow, comes into play. The loss of function may be temporary or permanent depending upon the intensity and duration of the inflammation. The intensity and duration of the inflammation is in turn dependent upon the agent inducing the inflammation and host's immunological capability to respond to the agent.

The eye is particulary susceptible to damage caused by inflammation due to its need for a clear medium. Any changes in this medium will interfere with the transmittance of light and be immediately perceived [1]. This is why it is important clinically to control inflammation in the eyes.

Ideally, the objective in the treatment of inflammation is to identify, control or avoid the injurious agent. However, most of the time this is not possible as in the cases of intraocular surgery and laser treatment of the eyes. Once the inflammatory process is set in motion, the chemical mediators of inflammation need to be identified and controlled. The difficulty lies in eliminating the damaging effects of inflammation while preserving its salutatory effects.

Chemical Mediators of Inflammation

The inflammatory process is complicated and is not yet fully understood. Some of the known mediators of inflammation can be classified as plasma origin and tissue origin. Mediators of plasma origin include the complement system, kinin system and clotting system. Mediators of tissue origin include vasoactive amines, arachidonic acid cascade, lysosomal components, lymphocyte products and lens proteins.

Complement Systems

The complement system consists of 18 bioactive peptides which function to increase vascular permeability, chemotaxis, opsonization, and finally lysis of cell membranes. The complement system is activated via two pathways. The classical pathway is initiated by antigen-antibody complexes. The alternate pathway is activated by nonimmunologic stimuli such as endotoxins nad IgG aggregates. The protein components having biologic activity include: (a) C3a, which increases vascular permeability by inducing release of histamines from mast cells; (b) C5a, which also increases vascular permeability by release of histamine; it also activates lipoxygenase in neutrophils and macrophages and is highly chemotactic to neutrophils and monocytes; (c) C3b, which is an opsonizing agent; (d) C567, a chemotactic agent, and (e) C5b9, the final lytic component [2, 3]. Some of the smaller peptide fragments such as C_4 and C_2 have kinin-like activity [4].

Kinin System

The kinin system is involved in the production of a vasoactive non-peptide called bradykinin. Bradykinin can cause pain, increased vascular permeability, smooth muscle contraction and dilatation of blood vessels [5]. However, bradykinin does not produce chemotaxis, but other components of the kinin system do possess chemotactic activity [3].

The kinin system begins with the activation of factor XII (F-XII), also known as Hageman factor, of the clotting system. F-XII is activated by contact with collagen, basement membrane or endotoxins. A fragment of F-XII is produced which converts prekallikrein to kallikrein. Kallikrein converts high molecular weight kininogen to bradykinin. Bradykinin itself is short lived, but it further activates F-XII and results in a self-amplification of the kinin system [5–7].

The Clotting System

In the final step of the clotting cascade is the production of fibrin from fibrinogen. During this reaction, reactive fibrin peptides are also formed. These peptides act as chemotactic agents and also produce an increase in vascular permeability.

The fibrinolytic portion of clotting system, plasmin, is produced from plasminogen either by kallikrein or plasminogen activator. Besides the lysis of clots, plasmin is capable of (a) activating F-XII which further generates bradykinin; (b) producing C3a, and (c) inducing the formation of fibrin split products which cause increased vascular permeability [6].

Vasoactive Amines (Histamine and Serotonin)

Histamine is present in mast cells, basophils, platelets, and the parietal region of the stomach. Serotonin (5-HT) is present in mast cells, platelets, intestinal mucosa and the brain. These vasoactive amines are released from these cells in response to physical injury, certain chemical agents, and immunological processes. Numerous other agents are also released from these cells along with histamine and 5-HT, particularly the slow-reacting substance of anaphylaxis (SRS-A) and platelet-activating factor (PAF), which are responsible for anaphylaxis. As for histamine and 5-HT, they are known to produce contraction of smooth muscle and increase vascular permeability. However, neither has any chemotactic effect on leukocytes.

Arachidonic Acid and Its Metabolite

The study of arachidonic acid, prostaglandins, and leukotrienes is rapidly expanding. The metabolites of arachidonic acid have a variety of biological functions and inflammation is only one of them.

Arachidonic acid is a 20-carbon polyunsaturated fatty acid which does not exist in a free form. Instead, it is esterified in cell membrane phospholipids. Arachidonic acid can be released by phospholipase and by various stimuli which include mechanical, chemical and other mediators of inflammation such as C5a.

Prostaglandins can be thought of as autocoids, short range hormones, which exert their actions locally and are rapidly degraded. The prostaglandin arm of arachidonic acid metabolism is initiated by cyclooxygenase to produce a prostaglandin endoperoxide, PGG_2, which is converted to PGH_2. Both of these are capable of stimulating platelet aggregation and contraction of smooth muscles [8, 9].

PGH_2 gives rise to the following products: (a) thromboxane A_2 (TXA_2); (b) prostacyclin (PGI_2), and (c) stable prostaglandins, PGE_2, $PGF_{2\alpha}$, PGD_2. TXA_2 causes platelet aggregation and blood vessel contraction [9]. PGI_2 antagonizes the effect of TXA_2, inhibiting platelet aggregation and inducing vasodilatation. PGE_2, $PGF_{2\alpha}$, and PGD_2 have a variety of actions on vascular tone and permeability. Their actions depend on the vascular bed and vary according to species [8–10].

The second arm of the arachidonic acid metabolism cascade is activated by lipoxygenase. The end products are leukotrienes. Lipoxygenase converts arachidonic acid to hypoperoxyeicosatetraenoic acid (HPETE). HPETE may either be converted to HETE, which is a potent chemotactic agent for neutrophils, or converted to leukotriene. The first leukotriene is LTA_4, which is an unstable compound. LTA_4 may give rise to either LTB_4 or LTC_4. LTB_4 is chemotactic and LTC_4 induces vasoconstriction. LTC_4 may give rise to LTD_4 and LTE_4 [11, 12]. Leukotrienes are closely related to inflammation processes particularly in the later phase.

Lysosomal Components

Lysosomal components include: (a) cationic proteins which increase vascular permeability and cause chemotaxis of mononuclear phagocytes; (b) acid proteases which degradate basement membranes, and (c) neutral proteases which degradate collagen, elastin, renal basement membrane, cartilage, fibrin, and kininogen [13, 14]. Release of these lysosomal components results in damage to tissue elements and perpetuation of the inflammation process.

Lymphocyte Products

In the cell-mediated immune reactions, antigens trigger sensitized T lymphocytes to release an array of mediators called lymphokines. These mediators cause inflammation-related events through the following agents:

(a) migration inhibitory factor; (b) chemotactic factor; (c) lymphotoxin; (d) skin-reactive factor; (e) mitogenic factors; (f) endogenous pyrogen, and (g) lymph node permeability factor [9, 15].

Ocular Inflammation

The inflammatory response in all tissues follows a stereotypic pattern, and the eyes are no exception. However, due to the unique anatomy and physiology of the eye, the inflammatory response is modified. Different portions of the eye have varying capabilities in mounting an inflammatory response.

Conjunctiva

The conjunctiva is highly vascularized and highly sensitive to a variety of agents. The reason for this is the presence of histamine (H_2) receptors in the conjunctival vessels and abundant number of plasma cells including mast cells [9]. Vasodilatation and congestion can be blocked by cimetadine (H_2-blocker) [16]. However, antihistamine appears to have little or no effect on intraocular inflammation [8, 17–19].

Cornea

The cornea is on the opposite end of the spectrum of inflammatory response. The cornea is said to be 'immunologically privileged', due to its avascularity. As a result, an agent which is capable of eliciting a strong and rapid response elsewhere might only elicit a slow and weak inflammatory response in the cornea. This is due to the fact that the agent is unable to diffuse in significant amount into the cornea to evoke a reaction. In addition, the chemical and cellular components of inflammation are restricted from reaching the injured site to produce a strong inflammatory response.

Lens

The lens of the eye is another immunologically privileged site. It is avascular like the cornea and is further isolated from the rest of the body by a capsule. Interestingly, it has been demonstrated that the lens is capable of producing prostaglandin E_2 (PGE_2), a potent mediator of inflammation [20–22]. The role of the lens during intraocular inflammation is uncertain at this point.

Iris and Ciliary Body

The iris and ciliary body are highly vascular. However, tight junctions are present in nonpigmented ciliary epithelium and in the vessels of the iris forming the blood-aqueous barrier. With inflammation and application of prostaglandins, the tight junction becomes leaky. It has been shown by electron microscopy that the ciliary epithelium is a source of protein leakage [23–26]. However, the iris shows remarkably little protein exudate [25, 27, 28]. The iris does play a very important role in inflammatory response of the anterior chamber. It was first noted by Ambache [29] that the iris produces a substance that induces an inflammatory response. The substance was termed irin. It was later shown that irin is a mixture of prostaglandins. The iris is able to produce and release prostaglandins after mechanical, chemical, thermal and radiation stimuli.

Retina and Choroid

Tight junctions also exist in retinal vasculature and in the retinal pigment epithelium. These represent the blood-retinal barrier and limit access of antibodies and leukocytes to the noxious agent. In contrast, the choroidal vasculature can be considered leaky. The vessels of the choroid are permeable enough to allow deposits of circulating antigen-antibody complexes which in turn set up an inflammatory response [30, 31].

Lens Protein-Induced Inflammation

It is well established that lens protein is capable of inducing a delayed inflammatory response [32–35]. The generally accepted theory is that lens proteins are sequestered during early fetal life and are hidden from the self-recognition process of the immunologic development [9, 18]. Any cause of lens protein leakage (trauma, hypermature cataract, or ocular surgery) may produce an inflammation response. Most of the studies to date have demonstrated that the inflammation induced is due to a hypersensitivity reaction [32–34], which requires prior sensitization.

Acute Lens Protein-Induced Inflammation

It has been demonstrated that lens protein [36] is also capable of producing an acute inflammation without prior sensitization. Whole lens protein extracts from white albino rabbits are injected intracamerally into other white albino rabbits. Within 2 h of the injection, the iris and ciliary body exhibit redness and congestion which will gradually subside after 48 h.

This ocular inflammatory response can be graded by two methods. The first method is done with slitlamp observation and a scoring system [36]. The second method, which has only recently been used for this model, is the measurement of fluorescein leakage by fluorophotometry. Albeit the results obtained with these two methods are similar, both methods have advantages and disadvantages.

The slitlamp observation is semiobjective. The results are based on the observer's interpretation and bias. However, this method provides a better overall picture of the inflammatory response of the eye by recording the redness of the iris and the cloudiness of the anterior chamber. While fluorophotometry is more objective and more sensitive, it only measures the amount of fluorescein leakage which correlates to the breakdown of the blood-aqueous barrier [36]. It is possible to have an intense leakage of fluorescein with minimal redness of the iris and a clear anterior chamber.

Mechanism of Lens Protein-Induced Inflammation

Attempts have been made to elucidate the mechanism of lens protein-induced inflammation. Cyclooxygenase inhibitor (indomethacin), lipoxygenase inhibitors (nordihydroguaiaretic acid (NDGA) and phenidone), and antihistamines (H_1 and H_2 blockers) were topically applied 1 h prior to lens protein injection. Not surprisingly, antihistamines, both H_1 and H_2, did not produce any significant reduction in the inflammatory response. Indomethacin produced maximal reduction of the inflammation within the first 2 h. NDGA and phenidone produced maximal reduction of inflammation after 3–4 h. This suggests that the acute inflammatory response can be divided into an early phase and a late phase. The early phase is mediated by prostaglandins and the late phase is mediated by leukotrienes [36]. Paradoxically, leukotriene is not detected by use of high-performance liquid chromatography. It is also made more difficult to conclude that leukotrienes are involved in the late phase of inflammation by reports that NDGA and phenidone both possess lipoxygenase and some cyclooxygenase inhibitory properties [38–40]. The question now arises as to whether leukotrienes have a role in the inflammation of the eye. The answer is probably yes.

Studies with either topical or intracameral application of prostaglandins [41] show that prostaglandins' action are limited to the production of redness of the iris, increase in protein exudate in the aqueous humor, and increase in intraocular pressure, but no leukocytes are observed [41, 42]. Without the cellular component of inflammation, the inflammatory process is not very damaging and is self-limiting.

It has been recently demonstrated that the anterior uvea of the monkey, dog, cat, rabbit, quinea pig, rat and human is capable of producing metabolites of arachidonic acid when incubated with lipoxygenase [43–45]. In addition, pretreatment of the iris with cyclooxygenase inhibitor appears to facilitate the production of leukotrienes [42]. Another possible source of leukotrienes is the lens. Bovine lens epithelial cells produce leukotrienes after incubation with arachidonic acid [46]. Presently it is unknown whether other species are capable of leukotrine production from the lens.

There are several possibilities to explain how the lens protein may initiate an inflammatory response in the eye. Lens protein may irritate the iris resulting in the release of prostaglandins and leukotrienes. Also the lens protein may contain prostaglandins and leukotrienes, since it appears that the lens is capable of producing these mediators. Another contributing factor is the chemotactic activity of the lens proteins. Lens proteins are capable of attracting both monocytes and polymorphonuclear leukocytes [47, 48].

So far, only the lipid mediators of inflammations have been discussed in the lens protein-induced inflammation. Another factor which may be involved is the neurogenic release of substance P. There is evidence that it is responsible for immediate increase in the vascular permeability in the eye after topical application of nitrogen mustard [49, 50]. Whether substance P plays a significant role in the inflammatory process remains to be determined.

Anti-Inflammatory Agents

There are presently four major classes of anti-inflammatory agents. They include cromolyn sodium, antihistamines, steroids and nonsteroidal anti-inflammatory drugs (NSAIDs). None of these drugs are ideal and only the steroids and NSAIDs are of any benefit in reducing intraocular inflammation.

The steroidal agents are the most effective in reducing inflammation, but are also the most nonspecific. Steroids act by entering the target cell and entering the cell's nucleus where it binds with the DNA [19, 51]. The target cell responds by increasing or decreasing the production of a specific protein. The net effect is very diverse depending on the target cell [15].

Steroids can reduce the products of cyclooxygenase and lipoxygenase. The target cell is stimulted to produce a protein (lipomodulin) which inhibits phospholipase productions [19, 52, 53]. Steroids can further reduce inflammation by reducing the release of vasoactive amines from mast cell and basophils [54]. The migration of macrophages is also reduced by steroids [55, 56].

The complications of steroidal therapy are numerous. The most noticeable side effect of steroid when applied topically to the eye is the possibility of causing an occular hypertensive crisis. The NSAIDs are not as toxic, but neither are they ideal. They may at times intensify inflammatory response. As two arms of the arachidonic acid cascade, the cyclooxygenase and lipoxygenase system are interrelated. If one system is blocked, more products of the other will be produced. All of the marketed NSAIDs produce their anti-inflammatory effect by blocking the cyclooxygenase system, although some may have some degree of inhibiton on the lipoxygenase [9]. By blocking the cyclooxygenase system, more substrate becomes available for the production of leukotrienes. This has been shown to cause a more intense accumulation of leukocytes in the aqueous humor which leads to more scarring [18, 57].

New Drugs for Lens Protein-Induced Inflammation

There is a need for a more specific NSAID. Most of the marketed NSAIDs are derivatives of salicylic acid, pyrozolin indole acetic acid, pyrolic acid, propionic acid, anthronilic acid and oxicam. These classes of NSAIDs do not have significantly differing anti-inflammatory properties [9]. It is doubtful that new breakthroughs can be accomplished by further modifying these structures.

Matrine

An area to search for new classes of NSAIDs is from traditional herbal medicines used in different cultures. One such new compound is matrine which is isolated from the root of *Sophora subprostrata.* This root has been used for centuries as an analgesic, antipyretic, and inti-inflammatory agent in Chinese medicine. Several studies have shown that matrine is the active agent of *S. subprostrata.*

Matrine's antipyretic property has been demonstrated by injecting yeast into rats to induce fever. Matrine, when given either orally or intra-

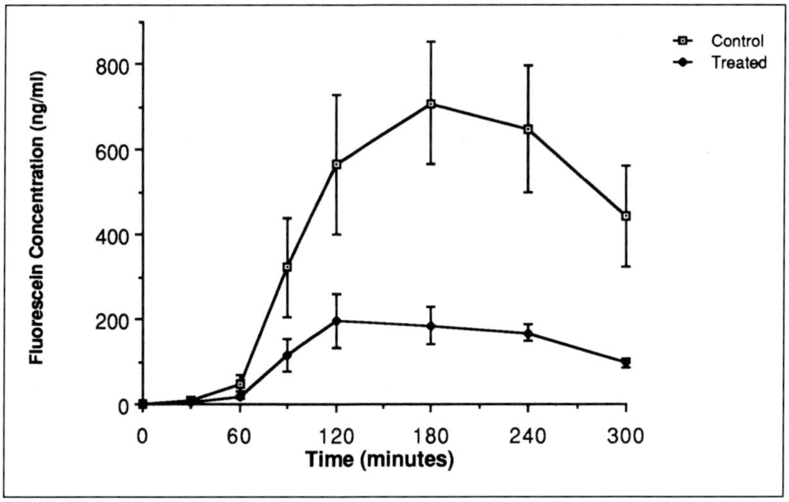

Fig. 1. Effects of 2% matrine on ocular inflammation induced by lens protein. The concentration of fluorescein in the anterior chamber of the eye is measured by fluorophotometry. Each point is a mean of six values and the bars represent SEM.

peritoneally, showed a dose-dependent reduction in rectal temperature [58]. Matrine's anti-inflammatory action is demonstrated in mice with carrageenan paw edema [59]. Within several hours after treatment with matrine, there is significant reduction of paw edema. Matrine's analgesic property has been demonstrated by prolonging the paw-lick reflex time of mice on a hot plate [60].

Recently, matrine has been shown to be effective in reducing lens protein-induced ocular inflammation after topical pretreatment. By using a scoring system [36], matrine produced maximal reduction in inflammation within 3–4 h and the reduction of inflammation which is evident up to 48 h [60]. Similar results are also obtained when fluorophotometry is used to measure fluorescein leakage (fig. 1). Matrine reduces fluorescein leakage throughout the 5-hour time course but maximal reduction of fluorescein leakage occurs at the end of the 5-hour period. The manner in which matrine reduces fluorescein leakage is very similar to NDGA and phenidone (fig. 2, 3). NDGA and phenidone are reported to possess both lipoxygenase and cyclooxygenase inhibitory properties. It is quite possible that matrine may also possess lipoxygenase and cyclooxygenase inhibitory properties. This hypothesis is further supported when comparing the time

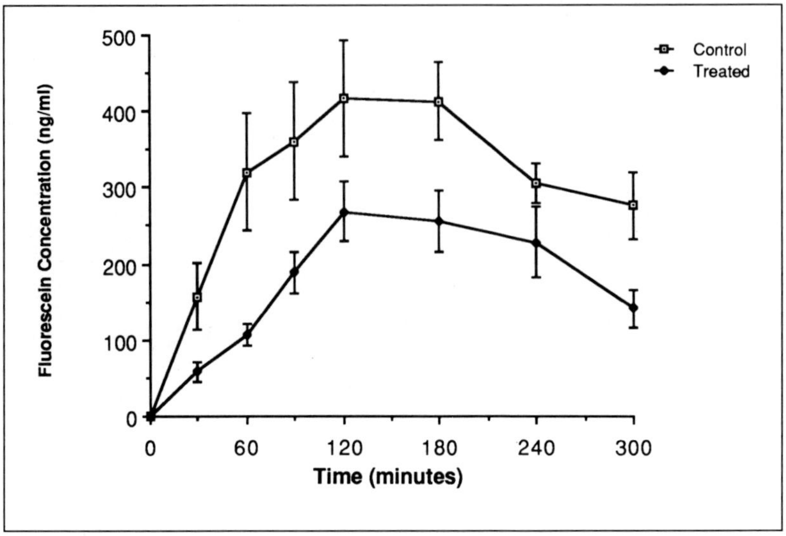

Fig. 2. Effects of 1% NDGA on ocular inflammation induced by lens protein. Each point is a mean of eight values and the bars represent SEM.

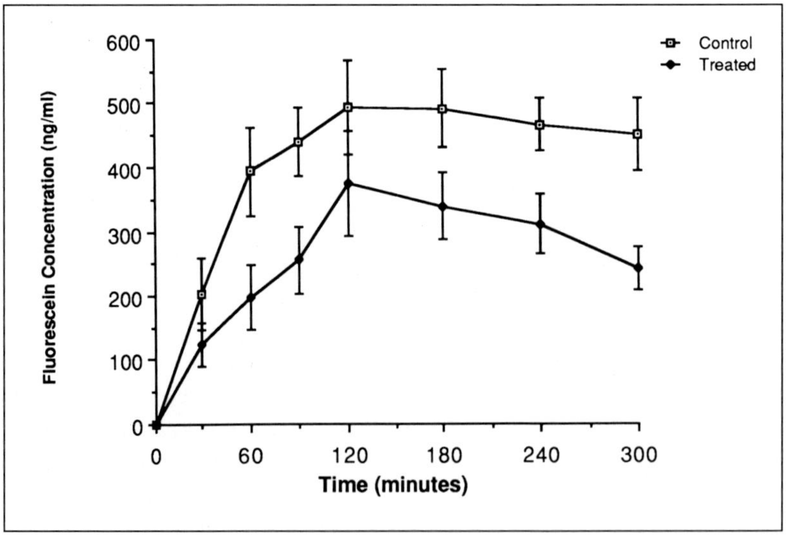

Fig. 3. Effects of 1% phenidone on ocular inflammation induced by lens protein. Each point is a mean of six values and the bars represent SEM.

of onset of matrine's maximal anti-inflammatory effect with that of a cyclooxygenase inhibitor, and a lipoxygenase inhibitor: matrine's onset of maximal activity is at the intermediate [36, 51, 60].

Lipoxygenase Inhibitors and Leukotriene Receptor Blockers

Presently there are no marketed NSAIDs which possess potent lipoxygenase inhibiting activity. A new lipoxygenase inhibitor has been recently tested with interesting results.

REV 5901 is a new potent lipoxygenase inhibitor. This compound has been shown to decrease inflammation more effectively than NDGA or phenidone when the slitlamp grading system is used to record the inflammation [51]. However, when fluorophotometry is used to measure fluorescein leakage, REV 5901 produced a rapid and intense leakage of fluorescein into the anterior chamber within the first hour after injection of lens protein. This was not observed with NDGA or phenidone (fig. 2, 3).

These observations may be explained by interrelationship between the two arms of the arachidonic acid cascade. Treatment with REV 5901 results in greater poduction of prostaglandins by strongly inibiting the lipoxygenase activity. This in turn leads to a more severe breakdown of the blood-aqueous barrier during inflammation as demonstrated by the high concentration of fluorescein in the anterior chamber. This intense initial leakage of fluorescein is not seen with NDGA or phenidone, and is most likely due to the lipoxygenase inhibitory activity. This indicates that a pure potent lipoxygenase is not a good anti-inflammatory agent to use on the eye alone, since strictly inhibiting lipoxygenase will lead to a higher concentration of prostaglandins. This will result in an increase in the intraocular pressure [61].

Recently, indomethacin (a cyclooxygenase inhibitor) has been tested by the lens protein-induced inflammation model and fluorophotometry. In the indomethacin pretreated eye, no fluorescein leakage is measured up to 3 h after injection of lens protein. After 3 h, the fluorescein concentration quickly rises to a level equal or greater than the control eyes. By pretreatment with a combination of indomethacin and REV 5901, fluorescein level is significantly reduced throughout the entire 5-hour period. The degree of reduction in inflammation as measured by fluorophotometry with this combination is greatest when compared to the other test anti-inflammatory agents, including prednisolone. The combination of indomethacin and REV 5901 is a most promising new anti-inflammatory agent for clinical use.

More specific agents to antagonize the effect of the leukotrienes might also be useful in controlling ocular inflammation. These agents would include the leukotriene receptor blockers but no lipoxygenase inhibitors. Presently there exist blockers of LTC_4, LTD_4 and LTE_4 receptors [62]. Blocks of leukotriene receptors have not been tested in ocular inflammation, but they may prove useful.

Conclusions

Ocular inflammation has been reviewed along with chemical mediators of inflammation in general as a background in order to understand the lens protein-induced ocular inflammation which is closely related to the trauma, surgical operation and laser treatment of the eyes. Attempts were made to elucidate the mechanism by which ocular inflammation is induced by lens proteins. Clearly, histamine is not involved. Although arachidonic cascades are involved they are not the sole answer to the ocular inflammation induced by lens proteins. For example, lens proteins are capable of attracting both monocytes and polymorphonuclear leukocytes. The actual mechanism by which ocular inflammation is induced requires further elucidation. Several agents have been found to prevent and/or inhibit the ocular inflammation induced by lens proteins. Matrine is a natural product isolated from a medicinal plant whereas others are synthetic compounds. These compounds are not corticosteroids yet they produce anti-inflammation as effective as prednisolone. Further studies on these new agents are warranted to elucidate their action mechanisms. Hopefully, a new class of nonsteroidal anti-inflammatory agents could be found, which may or may not be related to the arachidonic cascade.

References

1 Peiffer R: Ocular immunology and mechanisms of ocular inflammation. Vet Clin North Am 1980:10:281–302.
2 Muller-Eberhard H: Complement. Annu Rev Biochem 1975;44:697–724.
3 Vogt W: Activation, activities and pharmacologically active products of complement. Pharmacol Rev 1974;28:125–169.
4 Klemperer M, Rosen F, Donaldson V: A polypeptide derived from the second component of human complement (C2) which increases vascular permeability. Clin Invest 1984;48:44–51.

5 Movat H: Kinins and the kinin system as inflammatory mediators; in Weissman G, Glynn I, Houch J (eds): Chemical Messengers of Inflammatory Process, Handbook of Inflammation. Amsterdam, Elsevier/North-Holland, 1979, vol 1, pp 47–112.
6 Wiggins R, Cochrane G: Hageman factor and the contact activation system; in Weissman G, Glynn L, Houch J (eds): Chemical Messengers of Inflammation. Handbook of Inflammation. Amsterdam, Elsevier/North-Holland, 1979, vol 1, pp 179–196.
7 Spragg J: The plasma kinin-forming system; in Weissman G (ed): Mediators of Inflammation. New York, Plenum, 1974, pp 85–111.
8 Kuehl F, Egan R: Prostaglandins, arachidonic acid and inflammation. Science 1980; 210:978–984.
9 Smolin E, O'Connor G: Ocular Immunology, ed 2. Boston, Little Brown, 1986, pp 1–101.
10 Zurier R: Prostaglandins, inflammation, and asthma. Arch Intern Med 1974;133: 101–110.
11 Samuelson B: Leukotrienes: Mediators of immediate hypersensitivity reaction and inflammation. Science 1983;220:568–575.
12 Stjernschantz J: The leukotrienes. Med Biol 1984;62:215–230.
13 Weiman G, Pieter J: Molecular basis of gouty inflammation: interaction of monosodium urate crystals with lysosomes and liposomes. Nature 1972;240:167.
14 Weissman G, Smolen J, Korchak H: Release of inflammatory mediators from stimulated neutrophils. N Engl J Med 1980;303:27.
15 DeWeck A: The biology of lymphokines; in Yamamura Y, Tada T (eds): Progress in Immunology. New York, Academic Press, 1983, pp 307–314.
16 Abelson M, Udell I: H_2 receptors in the human ocular surface. Arch Ophthalmol 1981;99:302–304.
17 Eakins K, Bhattacherjee P: Histamine, prostaglandins and ocular inflammation. Exp Eye Res 1977;24:299–305.
18 Kaden A: Developmental implications of ocular pharmacology. Pharmacol Ther 1985;28:163–226.
19 Masuda K: Anti-inflammatory agents: nonsteroidal anti-inflammatory drugs; in Sears M (ed): Pharmacology of the Eye. Berlin, Springer, 1984, pp 539–551.
20 Belisle E, Fu S, Eng P, et al: Prostaglandin E levels in normal and cataractous rat lenses. Invest Ophthalmol Vis Sci 1982;22(suppl):31.
21 Belisle E, Eng P, Fu S: Levels of prostaglandin E in rat lens in culture. Invest Ophthalmol Vis Sci 1983;24(suppl):202.
22 Fleisher L, McGolan M: Endotoxin-induced ocular inflammation increases prostaglandin E_2 synthesis by rabbit lens. Exp Eye Res 1985;40:711–719.
23 Meyers R, Shabo A, Maxwell D: Effect of prostaglandin on the blood-aqueous barrier in the rabbit ciliary process. Prostaglandins 1975;9:167–173.
24 Pederson O: Electron microscopic studies on the blood-glucose barrier of prostaglandin-treated rabbit eye. I. Iridial and ciliary processes. Act Ophthalmol 1975;53: 685–698.
25 Shigekumi O: Effects of paracentesis on the blood-glucose barrier: a light and electron microscopic study on cynomolgus monkey. Invest Ophthalmol 1976;15:824–834.
26 Vegge T, Neufeld A, Sears M: Morphology of the breakdown of the blood-aqueous

barrier in the ciliary processes of the rabbit eye after prostaglandin E_2. Invest Ophthalmol 1975;14:33–36.
27 Pederson O: Electron microscopic studies on the blood-aqueous barrier of prostaglandin-treated rabbit eye. II. Iris. Acta Ophthalmol 1975;53:699–709.
28 Szalay J, Goldberg R, Klug R: The effect of prostaglandin on the iridial blood vessel permeability. Acta Ophthalmol 1976;54:731–742.
29 Ambache N: Irin, a smooth-muscle contracting substance present in rabbit iris. J Physiol 1955;129:65–66.
30 Gamble C, Aronson S, Brescia F: Experimental uveitis. Arch Ophthalmol 1970;84: 321.
31 Howes E, McKay D: Circulating immune complexes. Arch Ophthalmol 1975;93: 365.
32 Apple D, Mammalis N, Steinmetz R, et al: Phacoanaphylactic endophthalmitis associated with extracapsular cataract extraction and posterior chamber intraocular lens. Arch Ophthalmol 1984;102:1528–1532.
33 Goldschmidt L, Goldbaum M, Walker S, et al: The immune response of homologous lens crystallin. I. Antibody production after lens injury. J Immunol 1982;129:1652–1657.
34 Goldschmidt L, Goldbaum M, Walker S, et al: The immune response to homologous lens crystallin. II. A model of ocular inflammation involving age injuries at separate times. J Immunol 1982;129;1658–1662.
35 Raki A, Misra R, Morgan G: Immunopathology of the lens. III. Humoral and cellular immune response to autologous lens antigens and their roles in ocular inflammation. Br J Ophthalmol 1977;61:371–379.
36 Miyano K, Chiou G: Pharmacological prevention of ocular inflammation induced by lens proteins. Ophthalmic Res 1984;16:256–263.
37 Sanders D, et al: Quantitative assessment of postsurgical breakdown of the blood-aqueous barrier. Arch Ophthalmol 1983;101:131–133.
38 Blackwell G, Flower R: 1-Phenyl-3-pyrazolidone: an inhibitor of cyclooxygenase and lipoxygenase pathways in lung and platelets. Prostaglandins 1978;16:417–425.
39 Hamberg M: On the formation of thromboxane B_2 and 12-*L*-hydroxy-5,8,10,14-eicosatetraenoic acid in tissue from the quinea-pig. Biochem Biophys Acta 1976; 431:651–654.
40 Morris H, Piper P, Taylor G, et al: The role of arachidonate lipoxygenase in the release of SRS-A from quinea-pig chopped lung. Prostaglandins 1980;19:371–383.
41 Kulkarmi P, Srinivasan B: The effect of intravitreal and topical prostaglandins on intraocular inflammation. Invest Ophthalmol Vis Sci 1982;23:383–392.
42 Flach A, Eliason J: Topical prostaglandin E_2 effects on normal human intraocular pressure. J Ocul Pharmacol 1988;4:13–18.
43 Kulkarmi P, Srinivasam B: Synthesis of slow-reacting substance-like activity in rabbit conjunctiva and anterior uvea. Invest Ophthalmol Vis Sci 1983;24:1079–1085.
44 Kulkarni P, Rodriquez A, Srinivsan B: Human anterior uvea synthesizes lipoxygenase products from arachidonic acid. Invest Ophthalmol Vis Sci 1984;25:221–223.
45 Williams R, Bhattacherjee P, Eakins K: Biosynthesis of lipoxygenase products by ocular tissues. Exp Eye Res 1988;36:397–402.
46 Lonchampt M, Bonne C, Regnault F, et al: Evidence of leukotriene B_4 biosynthesis in epithelial lens cells. Prostaglandins Leukotrienes Med 1983;10:381–387.

47 Henley W, Okes S, Leopold I: Human cellular immunity to lens leukocyte migration inhibition by lens protein. Arch Intern Med 1974;134:684–688.
48 Rosenbaum T, Samples T, Segmour B, et al: Chemotactic activity of lens proteins and pathogenesis of phacolytic glaucoma. Arch Ophthalmol 1987;105:1582–1584.
49 Comras P, Bito L: The pathophysiological effects of nitrogen mustard on the rabbit. II. The inhibiton of the initial hypertensive phase by capsaicin and the apparent role of substance P. Invest Opthalmol Vis Sci 1980;19:423–428.
50 Bill A, et al: Substance P: release on trigeminal nerve stimulation, effects in the eye. Acta Physiol Scand 1979;106:371–373.
51 Chiou L, Chiou G: Ocular anti-inflammatory action of a lipoxygenase inhibitor in the rabbit. J Ocul Pharmacol 1985;1:383–389.
52 Floman N, Zor U: Mechanism of steorid action in ocular inflammation: inhibition of prostaglandin production. Invest Ophthalmol Vis Sci 1973;16:69–73.
53 Flower R, Blackwell G: Anti-inflammatory steroids induce biosynthesis of a phospholipase H_2 inhibitor which prevents prostaglandin generation. Nature 1979;279: 456–459.
54 Weissman G, Thomas L: Studies on lysosomes. II. The effect of cortisone on the release of acid hydrolases from a large granule fraction of rabbit liver induced by an excess of vitamin A. J Clin Invest 1963:42:661–669.
55 Fabey J, Gugre P, Munck A: Mechanisms of anti-inflammatory actions of glucocorticoids. Adv Inflam Res 1984;2:21–51.
56 Parrillo T, Fauci A: Mechanisms of glucocorticoid action on immune processor. Annu Rev Pharmacol Toxicol 1979:19:179–201.
57 Kulkarni P, Bhattacherjee P, Eakins K, et al: Anti-inflammatory effects of betamethasone, dexamethasone phosphate and indomethacin on rabbit ocular inflammation induced by bovine serum albumin. Curr Eye Res 1981;1:43–47.
58 Cho C, Chuang C, Chen C: Study of the antipyretic activity of matrine. A lupin alkaloid isolated from *Sophora subprostrata.* Planta Med 1986;5:343–345.
59 Cho C, Chuang C: Study of the anti-inflammatory action of matrine: an alkaloid isolated from *Sophora subprostrata.* IRCS Med Sci1986;14:441–442.
60 Chuang C, Xiao J, Chiou G: Ocular anti-inflammatory actions of matrine. J Ocul Pharmacol 1987;3:129–134.
61 Trzeciakowski J, Chiou F, Watanabe K, et al: Effects of prostaglandins on intraocular pressure recovery rate in rabbits. Prostaglandins 1985;29:497–510.
62 Fleisch T, Rinkema L, Whitesitt C, et al: Development of cysteinyl leukotriene receptor antagonists; in Lewis A, Ackerman N, Otterness I (eds): New Perspectives in Anti-Inflammatory Therapies. Advances in Inflammation Research. New York, Raven Press, 1988, vol 12, pp 173–189.

George C.Y. Chiou, PhD, Institute of Ocular Pharmacology,
Department of Medical Pharmacology and Toxicology,
Texas A & M University College of Medicine, College Station, TX 77843 (USA)

Bazan NG (ed): Lipid Mediators in Eye Inflammation.
New Trends Lipid Mediators Res. Basel, Karger, 1990, vol 5, pp 110–129

The Influence of Eicosapentaenoic Acid-Enriched Diet on the Inflammatory Response of the Rabbit Eye

N.L.J. Verbeij[a], *N.J. van Haeringen*[b]

[a] Department of Ophthalmology, Erasmus University, Eye Hospital, Rotterdam, and [b] Netherlands Ophthalmic Research Institute, Amsterdam, The Netherlands

Prostaglandins (PGs) and leukotrienes (LTs) play an important role in ocular inflammation [1–8]. PGs contribute to the formation of edema and erythema [9, 10] and play an important role in the response of the eye to injury [11, 12]. LTB_4 is a potent chemotactic agent which plays a role in the migration of leukocytes into sites of inflammation [13–15].

These inflammatory mediators are derived from the fatty acid arachidonic acid (AA) by metabolic conversion with the enzymes cyclooxygenase and lipoxygenase (fig. 1). AA is released from the phospholipids of the cell membrane in response to injury or inflammation by phospholipase A_2 [16].

In addition to AA, some other fatty acids such as eicosapentaenoic acid (EPA) and dihomo-γ-linolenic acid (DGHL) may also give rise to PG and LT formation. If PGs and LTs are synthesized from a fatty acid, each fatty acid has its own PG and LT counterpart with different chemical structure and biological behavior (fig. 1) [17].

The fatty acid composition of the diet determines which fatty acids are incorporated in the phospholipids of the cell membrane and in what quantity [18]. Considerable evidence is available that human populations whose dietary protein is mainly derived from fish are at low risk with regard to cardiovascular disease [19–24]. A fish-derived diet also has a beneficial influence on edema formation, systemic lupus erythematosus, immunologically induced arthritis and glomerulonephritis in experimental

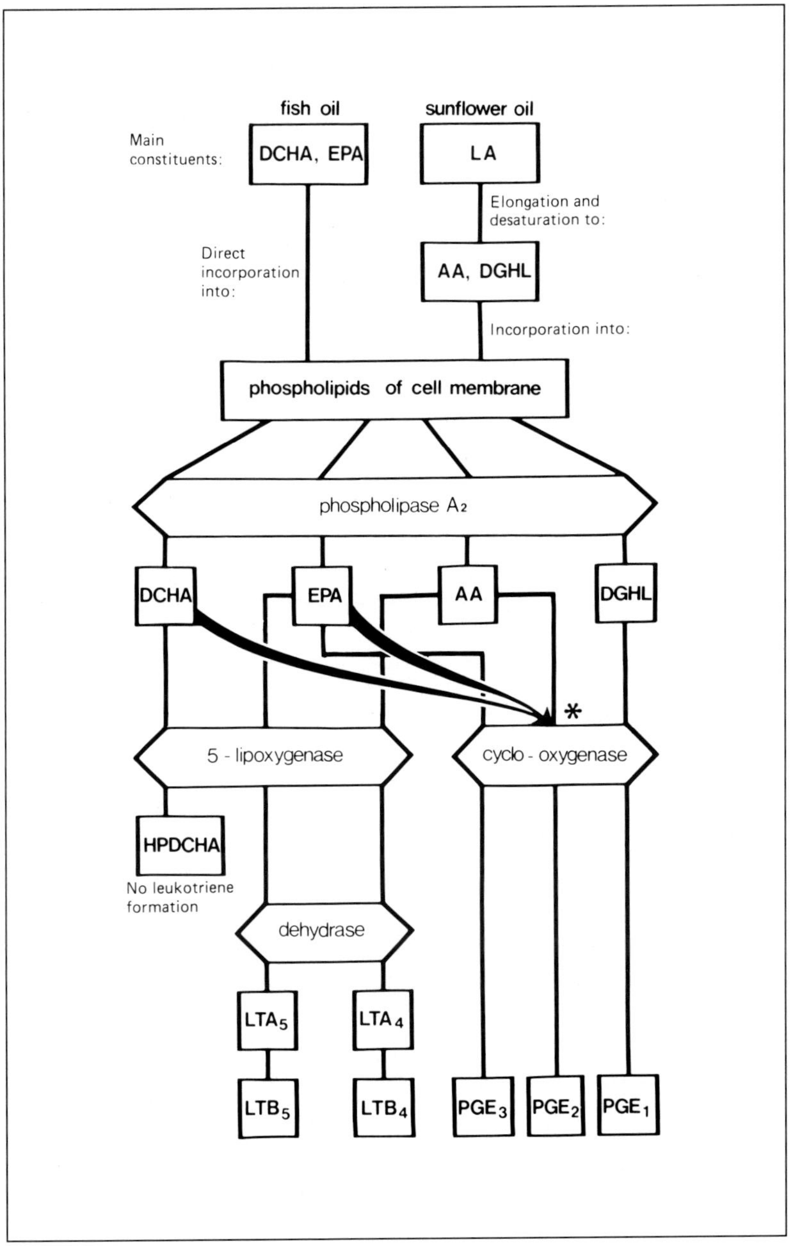

Fig. 1. Schematic representation of PGE and LTB formation from their precursor fatty acids. * Competitive inhibition.

animals [25–28]. It is believed that an important factor underlying the beneficial effect of a diet rich in seafood is the composition of the fish lipids, especially with reference to the essential polyunsaturated fatty acids that can serve as PG or LT precursors. Fish lipid contains large amounts of EPA which is only a very minor component of food derived from terrestrial plants and animals. EPA, present in food, leads to altered PG and LT synthesis [18, 29–32].

Competitive inhibition of metabolism of AA in the cyclooxygenase and lipoxygenase pathways by EPA and the formation of specific altered PGs and LTs (PGE_3 and LTB_5) from EPA may result from dietary enrichment with fish lipid (fig. 1) [17, 30–35].

We compared the influence of a fish oil diet with a sunflower oil diet, rich in linoleic acid (LA), on the injury response of the rabbit eye to laser coagulation. This injury response is normally accompanied by a breakdown of the blood-aqueous barrier and an increase in intraocular pressure (IOP) [2, 3, 5, 7]. We also compared the effect of these diets on the clinical appearance of human serum albumin (HSA)-induced immunogenic keratitis in rabbits. PGE and LTB formation by inflamed and noninflamed corneas was measured.

Materials and Methods

Animals

The experiments were performed in male pigmented chincilla rabbits weighing at least 2.0 kg. All eyes were initially examined with a slitlamp. Only animals without any sign of ocular inflammation were included in the study. During laser photocoagulation and paracentesis experiments the animals were sedated by intramuscular injection (0.75 ml/kg body weight) of Hypnorm® (Duphar, Amsterdam, The Netherlands) containing 10 mg fluanisone and 0.2 mg fentanyl base/ml. Local anesthesia was obtained by 0.2% oxybuprocaine eyedrops.

Diet and Feeding Procedures

The rabbits were maintained on standard rabbit diet (Hope Farms, Woerden, The Netherlands) prior to commencing the experimental diet. Because sufficient control of the exact fatty acid composition of a commercial rabbit diet is not possible, we have chosen a semisynthetic experimental diet of which the composition of nutrients can be exactly controlled. The fatty acid part was either sunflower seed oil (LA-enriched diet) or fish oil extracted from mackerel (EPA-enriched diet). The exact composition of the experimental diets is given in table 1.

The animals were gradually accustomed to their experimental diet in five steps of 2 weeks each. At each step a higher percentage of the standard diet was replaced by the experimental diet (either LA-enriched or EPA-enriched). Before the start of the experi-

Table 1. Composition of semisynthetic rabbit diet (28 energy % fat)

Ingredient	Fish oil diet g/1,000 kcal	Sunflower seed oil diet g/1,000 kcal	Percent in diet		Energy %	
Starch	140	140	43.7		49	
Casein	62	62	19.3		23	
Woody meal	60	60	18.7			
Salts mixture	25	25	7.8			
Vitamin mixture	2	2	0.6			
Grass meal	1.5	1.5	0.5			
Sulfaquinoxaline	0.04	0.04	0.0			
Fish oil	21.5	–	6.7	–	20	–
Sunflower seed oil	8.6	30.1	2.7	9.4	8	28
	320.6	320.6	100	100	100	100

ments, the animals were kept for at least 6 months on a 100% semisynthetic experimental diet (provided by Unilever Research Laboratories, Vlaardingen, The Netherlands). Diet and water were given ad libitum to both groups. Body weight was measured twice a week. The diet for the two groups of rabbits contained 30.1 g fat/1,000 kcal (energy % = 28, weight % = 9.4). In one group of rabbits the diet contained 9.4% sunflower seed oil (LA-enriched diet). In the other group the diet contained 2.7 weight % sunflower seed oil and 6.7% fish oil (EPA-enriched diet). The fish oil contained 13.3% EPA (20:5) and 7.1% docosahexaenoic acid (DCHA) (22:6).

Laser Photocoagulation of the Iris

Each rabbit received in one eye eight argon laser photocoagulation burns of 1,000 μm spot size made at 200 ms and 1,000 mW each. The contralateral eye served as a control for the IOP measurements. The laser applications were applied 2 mm from the pupillary margin, spaced circumferentially at regular intervals in a clockwise direction over 360°.

Intraocular Pressure

IOP measurements following a drop of 0.2% oxybuprocaine, were made with an Alcon pneumatograph (Alcon, Fort Worth, Tex.). The change in IOP in the laser-treated and control eye was calculated at each time interval by subtracting the baseline IOP from the measured IOP. The corrected change in IOP was then obtained by subtracting the change in IOP in the control eye from that in the laser-treated eye.

Protein Assay

Protein was measured according to Bradford [36] using HSA as standard.

Immunization

Immunization of the pigmented rabbits was performed by injection of 20 µl pyrogen-free HSA (20% solution; CLB, Amsterdam, The Netherlands) into the cornea of both eyes according to Morawiecki [37] after corneal anesthesia with 0.4% oxybuprocaine and sedation by Hypnorm® (fluanisone 10 mg/ml and fentanyl citrate 0.2 mg/ml) 0.75 ml/kg body weight.

Parameters of Ocular Inflammation

The effect of the diet on the clinical appearance of this HSA-induced immunogenic keratitis of the rabbit eye was evaluated by measuring corneal edema formation, neovascularization and the occurrence of an annular leukocyte infiltrate in the cornea (Wessely's ring) [38]. These three parameters of corneal inflammation can be well observed in vivo. The clinical observation was organized in a double-masked fashion.

Corneal Aspect. The number of days during which opaque rings or a diffuse completely opaque cornea were visible were counted from color slides taken daily (flash intensity 50 mW, distance 5 cm). For each animal the values of both eyes were averaged.

Neovascularization. The extent of vessel growth into the cornea was determined every other day from the color slides. Diapositives were projected and enlarged 12 times. By planimetry the area with new vessels was determined as a percentage of the total corneal surface. The percentage of vessel growth was plotted against time in days and from these graphs the area under the curve (AUC) was measured by planimetry. For each animal the values of both eyes were averaged.

Pachymetry. A Haag-Streit slitlamp with a pachymeter fitted with central fixation lights according to Mishima and Hedbys [39] was used for these measurements. The use of the fixation lights is essential in order to obtain consistent measurements of central thickness. From each eye the mean of three measurements was taken. For each animal the values of both eyes were averaged. Central corneal thickness was measured before intracorneal injection with HSA and from day 7 every other day until day 29. For each animal the difference between pachymetry measurement before intraocular injection with HSA and the measurement at any time thereafter was recorded as the change in corneal thickness. For each animal the corneal thickness was plotted in time and from this graph the AUC was measured.

Determination of Anti-HSA Immunoglobulins by ELISA

Serum was obtained from the rabbits of the two dietary regimens on day 29 after the first intracorneal injection with HSA. Microcuvettes (Gilford, Cleveland, Ohio) were coated with coating buffer containing 10 µg/ml of the HSA, used as antigen for the intracorneal injections on day zero. After a coating period of 1 h at room temperature the cuvettes were washed 3 times with phosphate-buffered saline (PBS) containing 0.1% Tween. Subsequently, 250 µl serum samples, diluted 1/10,000 in 0.1% Tween in PBS, were added to the cuvettes. After an incubation period of 1 h at room temperature the cuvettes were washed 3 times with 0.1% Tween in PBS. Subsequently, the cuvettes were incubated with a 1/500 dilution of peroxidase-conjugated goat anti-rabbit immunoglobulin for 1 h at room temperature (Nordic, Tilburg, The Netherlands). After being washed 3 times with 0.1% Tween in PBS, bound peroxidase-labelled antibody was developed at room temperature by adding 350 µl ABTS solution (0.16 m*M* 2,2-azino-di-3-ethyl-benzthiazoline-6-sulfonate; Boehringer Mannheim, FRG; 0.15% H_2O_2 in 0.05 *M* citric acid, pH 4.0). The green reaction product was measured after 20 min in a spectrophotometer (EIA; Gilford, Cleveland, Ohio) at 405 mm. As a standard a sample of purified rabbit anti-HSA immunoglobulin was used. This standard contained 10,000 ELISA units/ 0.37 μg of rabbit anti-HSA immunoglobulin.

Determination of Fatty Acids in Lipids of Corneal and Iris-Ciliary Body Tissue of the Two Dietary Groups

Eyes of 3 nonimmunized rabbits of each group of the two different dietary regimens were enucleated. The six dissected corneas or iris-ciliary bodies of rabbits of the same dietary regimen were pooled. One volume of methanol was added to the pooled samples and they were stored at –60 °C until biochemical analysis. Lipids were extracted from the corneal tissues with a mixture of chloroform:methanol (2:1). The chloroform layer was concentrated with a stream of nitrogen and the residue was transesterified with a mixture of hexane and diethyl ether (50:50) and evaporation of the solvent with a stream of nitrogen, the fatty acid methyl esters were purified over small silica columns with a mixture of hexane and diethyl ether (90:10). The fatty methyl esters were analyzed with gas-liquid chromatography after removal of the solvent with a stream of nitrogen.

A HP 5880 gas chromatograph equipped with an automatic sampler (7672 A, Hewlett-Packard) and a FID detector employing a WCOT glass capillary column (CP SIL 88; length = 25 m, inner diameter = 0.22) were used; injection temperature 255 °C, detection at 350 °C, programmed from 110 to 186 °C with 2 °C/min, and 10 min hold at the final temperature.

Calcium Ionophore Stimulated Production of PGEs and LTBs of Inflamed and Noninflamed Corneas in vitro

In 6 rabbits of each dietary regimen a second intracorneal injection of 20 μl of 20% HSA was given 30 days after the first injection. Three days after this second injection the animals were sacrificed with an overdose of pentobarbital and the eyes with inflamed corneas were enucleated. Also, eyes without inflammatory signs of the cornea were obtained from 6 rabbits of each dietary regimen, sacrificed 33 days after the first intracorneal injection.

For measurement of the stimulated production of PGE and LTB the 12 corneas of each group of animals were dissected and incubated at 37 °C in 2.0 ml of a HEPES-buffered solution (pH 7.4) of ionophore A23187, containing ionophore A23187 10 mg/l, HEPES 4.8 g/l, NaCl 8.0 g/l, $CaCl_2$ 0.2 g/l, KCl 0.2 g/l, $MgCl_2 \cdot 6H_2O$ 0.1 g/l and glucose 2.0 g/l. After 30 min, the incubation was stopped with 4.0 ml of cold methanol, acidified with 1 *M* acetic acid to pH 5.0 and the eicosanoids were extracted with 16 ml of chloroform, containing 50 ng ω-nor-PGE_2 and 50 ng of PGB_2 as internal standards. The organic layer was dried under a stream of nitrogen and the residue, taken up in a small volume of methanol, was divided into equal amounts on two silica columns (length 6 cm, internal diameter 0.6 cm) for analysis of PGEs respectively LTBs with high-performance liquid chromatography (HPLC).

Measurement of PGEs with HPLC

The less polar eicosanoids and other lipids were eluted from the silica column with a mixture of diethyl ether and acetic acid (100:0.5). The PG fraction was eluted with a mixture of diethyl ether, methanol and acetic acid (90:10:0.5). PGE was converted by alkaline treatment to PGB to facilitate the detection. Therefore, the PG fraction was dried under a stream of nitrogen and the residue was treated with 200 μl of 0.5 *N* aqueous KOH for 1 h at room temperature. The pH was adjusted to 5.0 with 1 *M* acetic acid and the PGBs were extracted twice with 2 ml of diethyl ether. The solvent was evaporated and the residue was rechromatographed on silica as described before; the PGBs were eluted with a mixture of diethyl ether and acetic acid (10:0.5). The fraction was evaporated under nitrogen and the residue was applied to the HPLC column in a small volume of methanol. The eluent was monitored at 278 nm. The amount of PGE was calculated from the relative peak areas in the chromatograph compared to that of the internal standard ω-nor-PGE_2.

Measurement of LTBs by HPLC

The neutral lipids and monohydroxylated fatty acids were eluted from the silica column with a mixture of hexane, diethyl ether and acetic acid (40:60:0.5). The LTB fraction was eluted with a mixture of ethanol and acetic acid (100:0.5). After evaporation of the solvent from the latter fraction with nitrogen, the sample was applied to the RP-HPLC column in a small volume of methanol. The eluent was monitored at 270 nm. The amount of LTBs was calculated from the relative peak areas in the chromatograph compared to that of the internal standard PGB_2, using the following molar extinctions: 270 nm LTB_4 = 50,000 and 270 nm PGB_2 = 23,700.

Chemicals and Equipment Used in HPLC

Uvasol grade chloroform and methanol were obtained from Merck. Baker-analyzed diethyl ether was distilled prior to use. Ultra pure water was generated from a Milli Q water purification system (Millipore Corp.), HPLC grade acetic acid was obtained from Baker. Chromatographic purification of the samples was achieved with silica columns (Kieselgel 60, 70–230 mesh ASTM; Merck). The internal standard ω-nor-PGE_2 was obtained enzymatically from the corresponding precursor fatty acid (19:4, 5, 8, 11, 14) [40]. PGB reference compounds and internal standard were obtained by alkaline conversion of the corresponding PGEs and isolated as described above for the isolation of PGEs from incubation mixtures with corneal tissue [40].

LTB_4 and LTB_5 reference compounds were obtained by incubation of human leukocytes with EPA during 20 min at pH 7.4 and 37 °C in the presence of ionophore A23187. The LTBs were isolated as described above for the incubation mixtures of corneal tissues.

A Waters associated model 510 A solvent delivery HPLC system, model U6K universal injector, and a variable wavelength Kratos model 773 UV detector were used in all determinations. Sample volumes of 10 μl were applied to the RP-18 column (Lichrospher 100 CH-18/2, 5 μm particle size, 250 × 4.6 mm).

Statistical Analysis

Data were analyzed by nonparametric methods to avoid assumptions about the distribution of the variables involved. Wilcoxon's signed-rank test was applied for pachymetry data obtained in the treated and untreated groups during the period of inflammation and the Mann-Whitney U test was applied for IOP data and served for analysis of the AUC of neovascularization and the duration of corneal opacification in the treated and untreated eyes at any given time. Significance of difference was obtained from a table for two-tailed observations, p values < 0.05 were regarded as significant. All values are given in mean ± SEM, except for the values of the PGEs and LTBs determined by HPLC and the data from fatty acid analysis. For these values no statistical analysis was performed.

Results

Iris Photocoagulation

The laser photocoagulation experiments were performed in 36 rabbits (18 on a fish oil diet and 18 on a sunflower oil diet). After 120 min aqueous was removed by paracentesis from both the laser-coagulated and uncoagulated eyes and used for protein and PGE_2 determination. In 6 animals of each group after 120 min after laser application no paracentesis was performed, but repeated IOP measurements were made. Three weeks after the laser photocoagulation experiments, paracentesis was performed in the 36 nonlaser-treated contralateral eyes of the same group of rabbits. We measured IOP immediately before and 15, 30, 60 and 120 min after laser treatment of the eye. In 24 rabbits 2 h after photocoagulation, anterior chamber paracentesis was performed. In the 12 other rabbits IOP was also measured 3, 4, 6, 24, 96 and 120 h after laser treatment.

Intraocular Pressure. Figure 2 illustrates the effect of dietary treatment on IOP at different time intervals after laser coagulation of the iris in 18 rabbits of each dietary group. In the sunflower oil fed group the IOP increased within 15 min after photocoagulation with a peak within 30 min; subsequently the IOP decreased below baseline value. In fish oil fed animals, the first 90 min after laser treatment IOP was significantly lower compared to sunflower oil fed animals. No increase in IOP occurred shortly after laser photocoagulation, even a small decrease in IOP was observed during the first hour in fish oil fed animals. After 90 min the course of the IOP was the same as in the sunflower oil fed animals.

Protein Assay. In 12 rabbits of each dietary group 2 h after iris photocoagulation, paracentesis of the anterior chamber was performed. The concentration of protein in the aqueous showed a significant difference between the rabbits on the sunflower oil diet and on the fish oil diet ($p < 0.01$) (table 2).

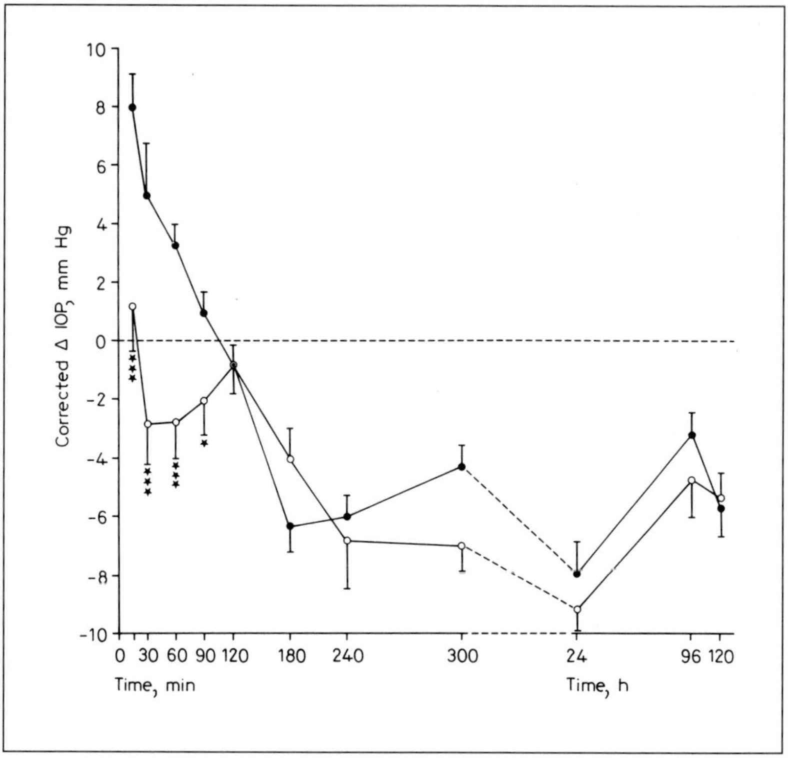

Fig. 2. The effect of laser photocoagulation of the pigmented rabbit iris on corrected change in IOP. The corrected change in IOP (mm Hg) is calculated by subtracting the change in IOP from baseline in the contralateral untreated eye from that in the laser-treated eye. IOP (mean ± SEM) was determined for laser treatment and contralateral eyes at 15 (n = 12), 30 (n = 12), 60 (n = 12), 120 (n = 12), 180 (n = 6), 240 (n = 6) and 360 (n = 6) min and on 24 (n = 6), 96 (n = 6) and 120 (n = 6) h for rabbits on an LA-enriched diet (●) and on an EPA-enriched diet (○). Significance of difference between the two dietary regimens was observed at 15, 30, 60 and 90 min after laser coagulation (* $p < 0.01$; *** $p < 0.001$).

Keratitis

The appearance of keratitis in rabbits on a diet of sunflower oil was as follows. One week to 10 days after intracorneal injection of HSA, clouding of the cornea started at the limbus and on about day 14–17 a white ring of opacification formed, also called Wessely's ring [38]. This ring was present for 1–8 days. Two to 4 days after Wessely's ring was first noted, the neo-

Table 2. Effect of the fatty acid composition diet on the concentration of protein in the anterior chamber of the rabbit eye after laser treatment

Treatment	n	Protein in aqueous humor, g/l	
		sunflower oil diet	fish oil diet
Laser	12	16.9 ± 1.1	10.9 ± 1.6[1]
Paracentesis prim. aqueous	18	0.4 ± 0.1	0.4 ± 0.1

[1] Significance of difference of protein values comparing the two dietary regimens ($p < 0.01$).

vascularization of the cornea started from the limbus. The neovascularization progressed until about day 22–25 and then regressed quickly. In all cases we observed a clear cornea 30 days after the injection of the HSA. All animals injected with HSA responded with white ring formation and neovascularization. Corneal edema formation, recorded with pachymetry, started around day 7 and lasted till day 30 (fig. 3, table 3).

In rabbits fed with the fish oil diet the period of corneal opacification was shorter in comparison with animals treated with sunflower oil diet (table 3). Also a white ring frequently appeared incomplete, and corneal opacification was less intense. Vessel growth measured as AUC was significantly diminished (table 3). A highly significant inhibition of corneal edema in fish oil fed animals was noted (table 3, fig. 3).

Systemic Antibody Response to Intracorneal Injected HSA

There was no significant difference in systemic response to HSA in the two different dietary regimens. Sunflower oil treated animals showed a serum level of 8.12 ± 2.17 μg/ml (mean ± SEM) of anti-HSA IgG immunoglobulins. The level in the fish oil treated animals was 8.38 ± 3.26 μg/ml serum (mean ± SEM).

Fatty Acids in Lipids of the Iris-Ciliary Body and Corneal Tissue

After 6 months on a fish oil diet, rabbits showed the occurrence of EPA and an increase of DCHA incorporated in the total lipid fraction of the iris-ciliary body and cornea compared to animals on a sunflower oil diet (table 4).

Table 3. Corneal opacity, growth of vessels and edema formation during immunogenic keratitis in rabbits on a diet with sunflower oil or fish oil

	n	Duration of corneal opacity days[1]	Vessel growth AUC, %[1]	Corneal thickness AUC, %[2]
Sunflower oil diet	10	4.8 ± 0.5	100 ± 15	100 ± 15
Fish oil diet	8	2.6 ± 0.6[a]	16 ± 8[b]	27 ± 6[c]

AUC = Area under the curve (100% in sunflower oil diet). All values are mean ± SEM. [a] $p < 0.01$; [b] $p < 0.001$; [c] $p < 0.005$.
[1] Significance of difference for duration of corneal opacity and vessel growth was calculated with the Mann-Whitney U test.
[2] Mean corneal thickness values of the animals on the two dietary regimens at various time points during the inflammation were tested for significance with Wilcoxon's signed-rank test.

Table 4. Effect of oral administration of sunflower oil or fish oil enriched diet on fatty acid composition of rabbit iris-ciliary body and corneal phospholipids

Numerical symbol	Trivial name	Fatty acid composition, mol%			
		iris-ciliary body		cornea	
		sunflower oil	fish oil	sunflower oil	fish oil
14:0		0.2	0.2	2.1	1.7
16:0	palmitic	14.0	16.0	14.0	14.0
18:0	stearic	19.0	18.0	9.0	8.8
18:1 (n–9)	oleic	11.0	12.0	37.0	38.0
18:2 (n–6)	LA	12.0	9.1	1.7	1.1
20:3 (n–6)		0.5	0.7	–	–
20:4 (n–6)	AA	19.0	16.0	4.3	3.4
20:5 (n–3)	EPA	0.0	2.6	–	0.5
22:4 (n–6)		3.3	0.8	1.2	0.2
22:5 (n–3)		0.0	2.7	0.1	0.7
22:6 (n–3)	DCHA	2.2	9.4	–	1.1

Values of six pooled iris-ciliary body and corneal preparations obtained from 3 animals, extracted and analyzed as described in the Methods section. The numerical symbol designates the chain length and the number of double bonds of fatty acid; n designates the place of the first double bond counted from the methyl end.

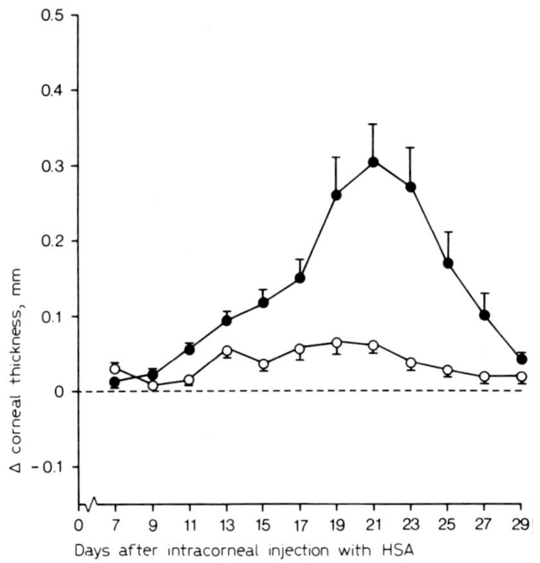

Fig. 3. Comparison of changes in mean corneal thickness during immunogenic keratitis treated with two different dietary regimens (fish oil diet (○), n = 8; sunflower oil diet (●), n = 10). Δ corneal thickness was calculated from the difference between the pachymetry measurement before intraocular injection with HSA and at any given time thereafter.

Calcium Ionophore A23187 Stimulated Production of PGEs and LTBs of Corneas in vitro

In uninflamed corneas of both dietary regimens, few eicosanoids were observed. In inflamed corneas of fish oil treated animals, significant production of PGE_2, LTB_4 and LTB_5 took place. The level of LTB_5 was 21% of the LTB_4 production. PGE_3 was produced in only minor amounts, 1.6% compared to the PGE_2 production. In inflamed corneas of sunflower oil seed fed animals, no LTB_5 or PGE_3 was produced. PGE_2 production was 33.5% and LTB_4 production was 31.5% higher compared to fish oil treated animals (table 5, fig. 4, 5).

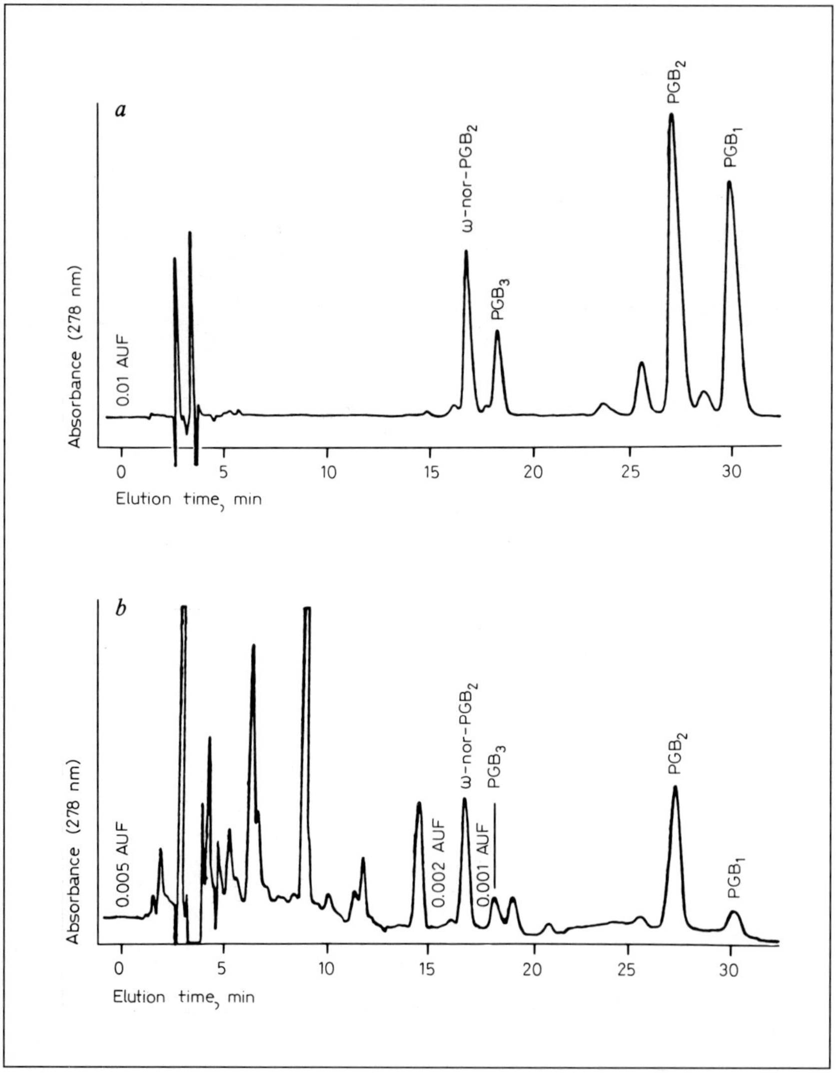

Fig. 4. HPLC chromatogram (*a*) of PGB reference compounds. Separation was achieved on a Lichrospher RP-18 (5), 250 × 4.6 mm, column with methanol, THF, water, acetic acid (50:12:38:0.1) at 1.0 ml/min. Representative chromatogram (*b*) showing the separation of PGE obtained after incubation of corneas from rabbits on a fish oil diet in the presence of calcium ionophore A23187. The PGEs produced were converted into the corresponding PGBs by alkaline treatment. The incubation mixture was processed according to procedures outlined under Methods. The absorption at 278 nm was recorded in various absorption units per full scale (AUF).

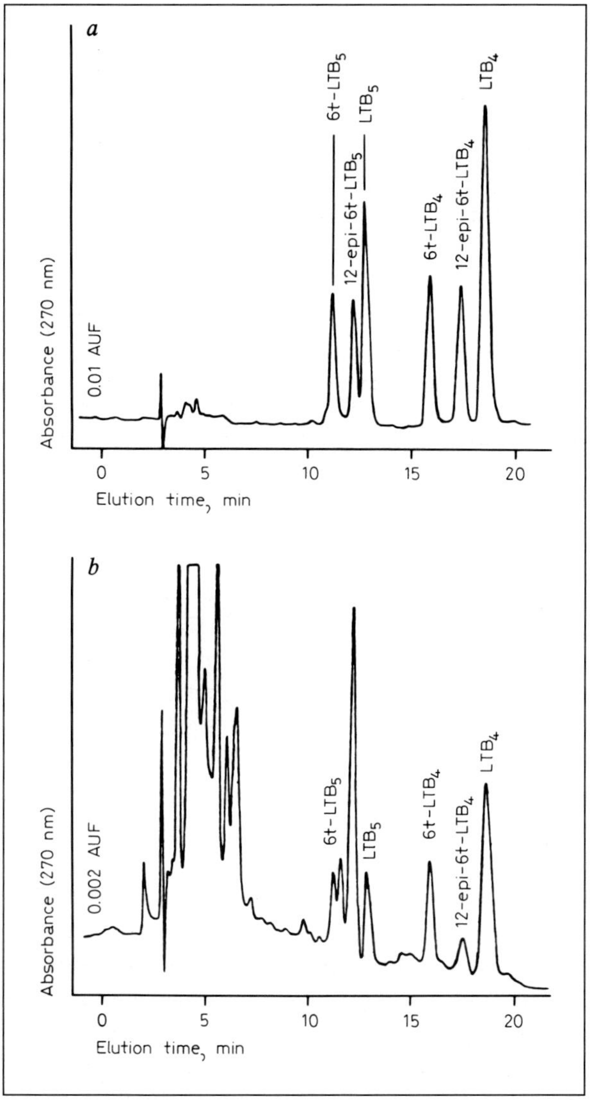

Fig. 5. HPLC chromatogram (*a*) of LTB reference compounds. Separation was achieved on a Lichrospher 100, CH-182, 250 × 4.6 mm, column with methanol, water, acetic acid (70:30:0.5) at 1.0 ml/min. Representative chromatogram (*b*) showing the separation of LTB obtained after incubation of corneas from rabbits on a fish oil diet in the presence of calcium ionophore A23187. The incubation mixture was processed according to procedures outlined under Methods. The absorption at 278 nm was recorded in various absorption units per full scale (AUF).

Table 5. Effect of dietary regimen on calcium ionophore A23187 stimulated production of PGEs and LTBs in corneal tissue of rabbits

	Fish oil diet		Sunflower oil diet	
	inflamed cornea	noninflamed cornea	inflamed cornea	noninflamed cornea
PGE_1	1.9	–	4.8	–
PGE_2	96.0	–	128.2	–
PGE_3	1.6	–	0.0	–
LTB_4	3.7	0.2	3.1	0.1
LTB_5	1.1	0.06	0.0	0.0

Values (ng/cornea) are calculated from RC-HPLC chromatograms of samples obtained, extracted and analyzed as described in the Methods section.

Discussion

In vitro cyclooxygenase inhibition can be achieved in different ways: by enzyme inhibition with nonsteroidal anti-inflammatory drugs or by competition with structurally related fatty acids of AA. Many different fatty acids can interact with the active site of cyclooxygenase, but the number and position of the double bonds determine how well inhibitors can prevent synthesis of PG.

Because the fatty acid composition change of a commercial standard rabbit diet is not possible in practice, we have chosen a semisynthetic experimental diet in which the composition of fatty acids could be regulated. We demonstrated an efficient incorporation of both EPA and DCHA from fish oil into the phospholipids of the iris-ciliary body with concomitant lower levels of AA. Phospholipase A_2 activation by a traumatic stimulus therefore may lead to a release of these polyunsaturated fatty acids besides lower levels of the naturally occurring AA from the cell membranes of the iris-ciliary tissue.

Inhibitors of PG synthesis have been demonstrated to reduce the disruption of the blood-aqueous barrier and the concomitant rise in IOP following argon laser after photocoagulation of the iris or puncture trauma of the cornea [12]. Likewise, competitive inhibition of PG synthesis by excess of foreign polyunsaturated fatty acids may in our experiments explain the stabilizing effect of fish oil feeding on the breakdown of the blood-aqueous barrier and on the IOP shortly after laser trauma.

For the study on the influence of a fish oil diet on ocular inflammation, we used a model of corneal anaphylaxis. The clinical effect observed with alteration of products of the 5-lipoxygenase and cyclooxygenase pathway confirms earlier observations on the involvement of PGs and LTs in symptoms of corneal inflammation [6–8].

The three parameters of corneal inflammation we used, Wesseley's ring, swelling and neovascularization, can be well observed in vivo and are known to be at least partially mediated by the formation of PGs and LTs [42–45]. The role of LTB_4 in infiltration of leukocytes and of PGE_1 and PGE_2 in neovascularization and edema formation of the cornea has been observed using these mediators implanted in polymer pellets in the rabbit cornea [43]. Suppression of these parameters in immunogenic corneal inflammation during the EPA-enriched diet can be regarded as evidence for altered formation of PGs or LTs. The effect of the fish oil diet was significant on all three clinical parameters. The EPA acid present in the fish oil diet may have several influences on an inflammatory response. It can act as a competitive inhibitor of cyclooxygenase [46, 47]. Also, formation of LTB_5 and PGE_3, eicosanoids which can only be synthesized from EPA, may take place [30, 32, 48–50]. If EPA gives rise to formation of LTB_5 and reduction in the formation of LTB_4, the effect results in reduced chemotactic activity. LTB_5 is at least 30 times less potent than LTB_4 in using aggregation chemokinesis and degranulation of polymorphonuclear leukocytes in vitro [51–53]. This in turn may lower leukocyte accumulation during the inflammatory response. The formation of PGE_3 from EPA and a reduced formation of PGE_2 and PGE_1 by competitive inhibition may lower PG-mediated symptoms like edema.

In addition, the presence of increased amounts of DCHA in fish lipids can alter PG and LT synthesis, however only by competitive inhibition of cyclooxygenase and to a lesser extent of lipoxygenase. PG and LT synthesis from DCHA is not possible (fig. 1) [54]. Therefore, the quantity of LTB and PGE production of the inflamed cornea in vitro is not necessarily a reflection of the in vivo production of the mediator. The effect of A23187 Ca ionophore on LT and PG formation in vitro is explained by its stimulation of phospholipase A_2 [55–57]. In vivo phospholipase A_2 is most probably the rate-limiting step in LT and PG synthesis [16, 58].

The proportionate production of PGE_1 and PGE_2 versus PGE_3 and the production of LTB_5 versus LTB_4 will reflect only the ratio of these inflammatory mediators in vivo. The in vitro A23187 Ca ionophore stimulated production of PGE_3 was only 1.6% of the total PGE level whereas

LTB_5 production was 20.3% of the total LTB production. This indicates that in vivo LTB_5 next to LTB_4 might play a role as mediator in the corneal inflammation of EPA fed rabbits, while the influence of EPA and DCHA on PG synthesis is mainly one of competitive inhibition with AA for cyclooxygenase. The formation of PGE_3 is of minor importance. After 6 months of fish oil diet, fatty acid analysis revealed the presence of EPA and DCHA in the cornea of the animals.

The LTB_5 produced by the noninflamed cornea is derived from EPA incorporated in the cornea itself. The greater production of LTB_5 in the inflamed cornea also may be due to production of LTB_5 from infiltrated leukocytes. In immunogenic keratitis the reaction in the cornea starts with the formation of antibody-antigen complexes. Although an effect of the EPA-enriched diet on antibody formation has been observed [59], in our experiments EPA feeding does not have an effect on the systemic antibody response to HSA.

In conclusion, EPA feeding has a significant therapeutic effect on laser-induced blood-aqueous barrier breakdown and IOP rise and on an immune complex inflammation of the rabbit cornea. The effect might be explained by competitive inhibition by EPA and DCHA of AA release from the phospholipids and by the formation of LTB_5 from EPA instead of the normal LTB_4 formation from AA.

Acknowledgements

We thank Mr. W. Heerooms for assistance with animal experiments, Mr. P. Moret and Mr. G. Kivits (Unilever Research Laboratory, Vlaardingen, The Netherlands) for the fatty acid analysis, PGE and LTB determinations, Unilever, Vlaardingen, for the animal diets, Mrs. M. Wissing and Mrs. T. Muijlwijk for secretarial assistance.

References

1 Eakins KE, Whitelocke RAF, Perkins ES, et al: Release of prostaglandins in ocular inflammation in the rabbit. Nature New Biology 1972;239:248–249.
2 Miller JD, Eakins KE, Atwal M: The release of PGE_2-like activity into aqueous humor after paracentesis and its prevention by aspirin. Invest Ophthalmol 1973;12: 939–942.
3 Bhattacherjee P, Hammond B, Salmon JA, et al: Chemotactic response to some arachidonic acid lipoxygenase products in the rabbit eye. Eur J Pharmacol 1981;73: 21–28.
4 Stjernschantz J, Sherk T, Borgeat P, et al: Intraocular effects of lipoxygenase pathway products in arachidonic acid metabolism. Acta Ophthalmol 1984;62:104–111.

5 Borodic G, Conte J, Aswad M, et al: Aqueous humor leukotriene levels in animal models of inflammation. ARVO Abstracts. Invest Ophthalmol Vis Sci 1985; 26(suppl):99.

6 Bazan HEP, Birkle DL, Beuerman R, et al: Cryogenic lesion alters the metabolism of arachidonic acid in rabbit cornea layers. Invest Ophthalmol Vis Sci 1985;26:474–480.

7 Bazan HEP, Birkle DL, Beuerman RW, et al: Inflammation-induced stimulation of the synthesis of prostaglandins and lipoxygenase reaction products in rabbit cornea. Curr Eye Res 1985;4:175–179.

8 Leibowitz HM, Ryan WJ, Kupferman A: The role of leukotriene B_4 in PMN chemotaxis in the inflamed cornea. ARVO Abstracts. Invest Ophthalmol Vis Sci 1985; 26(suppl):320.

9 Williams TJ: Prostaglandin E_2, prostaglandin I_2 and the vascular changes in inflammation. Br J Pharmacol 1979;65:517–520.

10 Moncada S, Ferreira SH, Vane JR: Prostaglandins, aspirin-like drugs and the edema of inflammation. Nature 1973;246:217–218.

11 Unger W, Cole DF, Hammond B: Disruption of the blood-aqueous barrier following paracentesis in the rabbit. Exp Eye Res 1975;20:255–270.

12 Unger WG, Perkins EG, Bass MS: The response of the rabbit eye to laser irradiation of the iris. Exp Eye Res 1974;19:367–377.

13 Ford-Hutchinson AW, Bray MA, Dois MV: Leukotriene B_4, a potent chemokinetic and aggregating substance released from polymorphonuclear leukocytes. Nature 1980;286:264–265.

14 Ford-Hutchinson AW: Leukotrienes: their formation and role as inflammatory mediators. Fed Proc 1985;44:25–29.

15 Higgs GA, Salmon JA, Spagne JA: The inflammatory effects of hydroperoxy and hydroxy acid products of arachidonate lipoxygenase in rabbit skin. Br J Pharmacol 1981;74:429–433.

16 Van den Bosch H: Intracellular phospholipases A. Biochim Biophys Acta 1980;604: 191–246.

17 Horrobin DF: The regulation of prostaglandin biosynthesis by the manipulation of essential fatty acid metabolism. Rev Pure Appl Pharmacol Sci 1983;4:339–383.

18 Schacky C, Fischer S, Weber PC: Long-term effects of dietary marine ω-3 fatty acids upon plasma and cellular lipids, platelet function and eicosanoid formation in humans. J Clin Invest 1985;76:1626–1631.

19 Dyerberg J, Bang HO, Stofferson E, et al: Eicosapentaenoic acid and prevention of thrombosis and artherosclerosis. Lancet 1978;ii:117–119.

20 Editorial: Dietary fats and coronary heart disease. Lancet 1980;i:414–415.

21 Culp BR, Lands WEM, Lucchesi BR, et al: The effect of dietary supplementation of fish oil on experimental myocardial infarction. Prostaglandins 1980;20:1021–1030.

22 Lorenz R, Spengler U, Fischer S, et al: Platelet function, thromboxane formation and blood pressure control during supplementation of the western diet with cod liver oil. Circulation 1983;63:504–511.

23 Kromhout D, Bosschieter EB, De Lezenne Coulander C: The inverse relation between fish consumption and 20-year mortality from coronary heart disease. N Engl J Med 1985;312:1205–1209.

24 Editorial: Fish, fatty acids and human health. N Engl J Med 1985;312:1253–1254.
25 Terano T, Salmon JA, Higgs GA, et al: Eicosapentaenoic acid as a modulator of inflammation. Biochem Pharmacol 1986;35:779–785.
26 Robinson DR, Prickett JD, Polisson R, et al: The protective effect of dietary fish oil on marine lupus. Prostaglandins 1985;30:51–75.
27 Prickett JD, Trentham DE, Robinson DR: Dietary fish oil augments the indication of arthritis in rats immunized with type II collagen. J Immunol 1984;132:725–729.
28 Prickett JD, Robinson DR, Steinberg AD: Dietary enrichment with the polyunsaturated fatty acid eicosapentaenoic acid prevents proteinuria and prolongs survival in NZB × NZW F_1 mice. J Clin Invest 1981;68:556–559.
29 Terano T, Salmon AT, Moncada S: Effect of orally administered eicosapentaenoic acid (EPA) on the formation of leukotriene B_4 and leukotriene B_5 by rat leukocytes. Biochem Pharmacol 1984;33:3071–3076.
30 Prescott SM, Zimmerman GA, Morrison AR: The effects of a diet rich in fish oil on human neutrophils: identification of leukotriene B_5 as a metabolite. Prostaglandins 1985;30:209–227.
31 Strasser T, Fischer S, Weber PC: Leukotriene B_5 is formed in human neutrophils after dietary supplementation with eicosapentaenoic acid. Proc Natl Acad Sci USA 1985;82:1540–1543.
32 Lee TH, Hoover RL, Williams JD, et al: Effect of dietary enrichment with eicosapentaenoic and docosahexaenoic acids on in vitro neutrophil and monocyte leukotriene generation and neutrophil function. N Engl J Med 1985;312:1217–1224.
33 Needleman P, Raz A, Minkes MS, et al: Triene prostaglandins: Prostacyclin and thromboxane biosynthesis and unique biological properties. Proc Natl Acad Sci USA 1979;76:944–948.
34 Hamberg M: Transformation of 5,8,11,14,17-eicosapentaenoic acid in human platelets. Biochim Biophys Acta 1980;618:389–398.
35 Needleman P, Whithaker MO, Wyche A, et al: Manipulation of platelet aggregation by prostaglandins and their fatty acid supervisors: Pharmacological basis for a therapeutic approach. Prostaglandins 1980;19:165–181.
36 Bradford MM: A rapid and sensitive method for the quantitation of microgram quantities of protein utilizing the principle of protein-dye binding. Anal Biochem 1976;72:248–254.
37 Morawiecki J: Präzipitationserscheinungen in die lebende Hornhaut bei Antigen-Antikörperreaktionen. Ophthalmologica 1956;132:236–243.
38 Wessely K: Über anaphylaktische Erscheinungen an der Hornhaut (experimentelle Erzeugung einer parachymatösen Keratitis durch artfremdes Serum). Münch Med Wochenschr 1911;58:1713–1714.
39 Mishima S, Hedbys BO: Measurement of corneal thickness with the Haag-Streit pachometer. Arch Ophthalmol 1968;80:710–713.
40 Kivits GAA, Nugteren DH: The urinary excretion of prostaglandins E and their corresponding tetranor metabolites by rats fed a diet rich in eicosapentaenoate. Biochim Biophys Acta 1988;958:289–299.
41 Unger WG, Bass MS: Prostaglandin and nerve-indicated response of the rabbit eye to argon laser irradiation of the iris. Ophthalmologica 1977;175:153–158.
42 BenEzra D: Neovasculogenic ability of prostaglandins, growth factors, and synthetic chemoattractants. Am J Ophthalmol 1978;86:455–461.

43 Rochels R: Tierexperimentelle Untersuchungen zur Rolle von Entzündungsmediatoren bei der Hornhautneovaskularisation. Doc Ophthalmol 1984;57:215–262.
44 Harvey P, Cehrry PMH: Indomethacin and dexamethasone in the suppression of corneal neovascularization. Can J Ophthalmol 1983;18:293–295.
45 Frucht J, Zauberman P: Topical indomethacin effect on neovascularization of the cornea and on prostaglandin E_2 levels. Br J Ophthalmol 1984;68:656–659.
46 Hornstra G, Haddeman E, Kloeze J, et al: Dietary fat-induced changes in the formation of prostanoids of the 2 and 3 series in relation to arterial thrombosis (rat) and atherosclerosis (rabbit). Adv Prostaglandin Thromboxane Leukotriene Res 1983;12:193–202.
47 Hornstra G, Christ-Hazelhof E, Haddeman E, et al: Fish oil feedings lowers thromboxane and prostacyclin production by rat platelets and aorta and does not result in the formation of prostaglandin I_3. Prostaglandins 1981;21:727–738.
48 Fischer S, Weber PC: Prostaglandin I_3 is formed in vivo in man after dietary eicosapentaenoic acid. Nature 1984;307:165–168.
49 Lee TH, Mencia-Huerta, Shih C, et al: Effects of exogenous arachidonic, eicosapentaenoic and docosahexaenoic acids on the generation of 5-lipoxygenase pathway products by ionophore-activated human neutrophils. J Clin Invest 1984;74:1922–1933.
50 Prescott SM: The effect of eicosapentaenoic acid on leukotriene B production by human neutrophils. J Biol Chem 1984;259:7615–7621.
51 Goldman DW, Pickett WC, Goetzl EJ: Human neutrophile chemotactic and degranulating activities of leukotriene B_5 derived from eicosapentaenoic acid. Biochem Biophys Res Commun 1983;117:282–290.
52 Terano T, Salmon JA, Moncada S: Biosynthesis and biological activity of leukotriene B_5. Prostaglandins 1984;27:217–228.
53 Lee TH, Mencia-Huerta JM, Shih C, et al: Characterization and biological properties of 5,12-dihydroxy derivatives of eicosapentaenoic acid, including leukotriene B_5 and the double lipoxygenase product. J Biol Chem 1984;259:2383–2388.
54 Corey EJ, Shih C, Cashman JR: Docosahexaenoic acid is a strong inhibitor of prostaglandin but not leukotriene biosynthesis. Proc Natl Acad Sci USA 1983;80:3581–3586.
55 Stenson WF, Parker CW: Metabolism of arachidonic acid in ionophore-stimulated neutrophils. J Clin Invest 1979;64:1457–1465.
56 Borgeat P, Samuelson B: Arachidonic acid and metabolism in polymorphonuclear leukocytes: Effects of ionophore A23187. Proc Natl Acad Sci USA 1979;76:2148–2152.
57 Knapp HR, Oelz O, Jackson Roberts L, et al: Ionophores stimulate prostaglandin and thromboxane biosynthesis. Proc Natl Acad Sci USA 1977;74:4251–4255.
58 Irvine RF: How is the level of free arachidonic acid controlled in mammalian cells? Biochem J 1982;204:3–16.
59 Prickett JD, Robinson DR, Bloch KJ: Enhanced production of IgE and IgG antibodies associated with a diet enriched in eicosapentaenoic acid. Immunology 1982; 46:819–826.

N.L.J. Verbeij, MD, Department of Ophthalmology, Erasmus University,
Eye Hospital, Schiedamsevest 180, NL–3000 LM Rotterdam (The Netherlands)

Bazan NG (ed): Lipid Mediators in Eye Inflammation.
New Trends Lipid Mediators Res. Basel, Karger, 1990, vol 5, pp 130–148

Platelet-Activating Factor and Cytokines Mediate and Amplify the Inflammatory Response at Site of Microvascular Injury[1]

Federico Bussolino, Alessandra Boles Carenini, Bruno Boles Carenini, Paolo Arese

Dipartimento di Genetica, Biologica e Chimica Medica, e Clinica Oculistica, Università di Torino, Italia

Impairment and modifications of the vessel wall functions characterize atherosclerosis, acute and chronic vasculitis, thrombosis and angiogenesis in course of tumor growth and in a variety of neovascularization processes in peripheral organs (in particular eye). Yet in the past decade, cellular and molecular approaches to the study of the biology of the vasculature have provided numerous insights into the pathogenesis of vascular damages. One of these is a response of the vascular wall to injury, having many of the features of the inflammatory response [1–5].

In vascularized tissues, the enusing inflammatory response may result in repair and healing or may initiate a process of chronic injury. Thus, dependent on various factors acting either individually or in combination, vascular injury may be reversible or initiate a process of scar formation leading to sclerosis and loss of function. In addition, an altered process of angiogenesis and production of extracellular matrix are the hallmarks of the progressive and chronic diseases. These may depend on: (1) an initial, massive and acute injury; (2) the persistence of the initiating pathogenetic mechanism that influences vascular cells directly or indirectly through the release of inflammatory mediators; (3) an altered regulation of the behav-

[1] This work was supported in part by Laboratori Decalande SpA, Italia.

ior of vascular cells triggered by the initiating inflammatory stimulus which persists despite the disappearance of the pathogenetic stimulus.

The initiating insult may determine an increased expression of gene coding for the synthesis of peptide growth factors or of matrix proteins acting on the cellular component of the vasculature with a modification of the 'vascular genetic program', that maintain an altered behavior of the vessel wall. This modified homeostasis is triggered by an increased production of autacoids (prostanoids, complement fragments, kinins, platelet-activating factor, vasoactive amines) and of cytokines that are the constitutive part of the so-called 'polypetide mediator network' which defines the network of interacting signals (interleukin-1, -2, -3, -4, -5, -6, -7, -8, interferons, colony-stimulating factors, tumor necrosis factors, transforming growth factors, platelet-derived growth factor, epidermal growth factor, fibroblast growth factor) that regulate growth, differentiation and function of cells involved in immunity, inflammation and hematopoiesis.

This review will focus on the role of certain cytokines and platelet-activating factor (PAF), a lipid mediator of cell-to-cell interaction and of inflammation [reviewed in 6] in the bidirectional interaction between leukocytes and endothelium, in determining and maintaining the vascular injury.

Cytokine Network Modulates Vascular Cell Functions

Inflammatory reactions and immunity involve strict interaction between immunocompetent cells and the vessel wall [7]. The extravasation and localization at inflammatory sites of leukocytes involve adhesion to and passage through endothelial linings. Changes in the vessel wall, including vasodilatation and proliferation of capillary endothelial cells (EC) and vascular smooth muscle, are well documented at sites of cell-mediated immune reactions involving leukocyte and platelet recruitment. Soluble products released by circulating cells (T cells, monocytes, neutrophils, platelets) are potent regulators of various functions of vascular cells, such as proliferation, migration, expression of class II histocompatibility antigens, changes of surface properties, production of lipid and peptidic mediators. The molecules which mediate the symbiotic relationship of circulating cells with vasculature have been, to some extent, defined recently and are summarized in table 1.

Table 1. Cytokine-inducible vascular cell effects

Effect	Cytokine
Endothelial cells	
Induction of class II MHC antigens [8]	INF-γ
Modulation of class I MHC antigens [9]	INF-α/γ, TNF-α/β
Induction of morphologic changes [10, 11]	INF-γ, IL-1, TNF-α
Modulation of leukocyte adhesion [12–16]	IL-1, TNF-α/β, INF-γ, TGF-β
Synthesis and release of mediators [17–25]	IL-1, TNF-α/β, IL-2
Modulation of hemostatic properties [17, 26–32]	IL-1, TNF-α, FGF, G-CSF
Modulation of migration and proliferation [10, 33–37]	TNF-α, G-CSF, GM-CSF, FGF, IL-1, INF-γ, TGF-α/β
Changes in matrix composition [10, 38]	TNF-α, IL-1, INF-γ
Smooth Muscle	
Modulation of proliferation [39–42]	PDGF, IL-1, TFG-β, INF-γ
Induction of class II MHC antigens [41]	INF-γ
Modification of intracellular ion concentration [43]	PDGF
Synthesis of IL-1 and TNF [44]	TNF-α

IL-1 – Interleukin-1; TNF = tumor necrosis factor; INF = interferon; TGF = transforming growth factor; FGF = fibroblast growth factor; G-CSF = granulocyte colony-stimulating factor; GM-CSF = granulocyte-macrophage colony-stimulating factor; PDGF = platelet-derived growth factor; MCH = major histocompatibility complex.

Interleukin-1 (IL-1) and Tumor Necrosis Factor (TNF) Play a Pivotal Role in Microvascular Immune Injury

The functional properties of vascular cells are profoundly affected by the inflammatory monokines IL-1 and TNFα/β. A first crucial aspect of EC function modulated by IL-1 and TNFα is the control of the movement of neutrophils and monocytes into inflammatory lesions or the homing of lymphocytes to lymphoid organs. This migration begins with the attachment of circulating cells to the vessel walls mediated by several families of adhesion molecules, including members of the integrin (CD18 complex) and immunoglobulin (intercellular molecular adhesion-1) supergene families, of the lectin, EGF, complement-cellular adhesion molecules (LEC-CAM) (endothelial-leukocytes adhesion molecule-1, granule membrane protein 140, Mel-14 antigen), and of Hermes antigens [reviewed in 45, 46].

IL-1 and TFNα increase the adhesion of leukocytes (neutrophils, monocytes, B and T cells) to endothelial cells. The effect of IL-1 is restricted to EC [12–14], whereas TNFα acts also on neutrophils [47]. Both monokines act by up-regulation of the intercellular adhesion molecule-1 (ICAM-1) [14] that is constitutive in EC and by the induction of the endothelial-leukocytes adhesion molecule-1 (ELAM-1) [13, 14]. ELAM-1 induction begins after 1–2 h, reaches the maximum after 4–6 h, declines to the basal level by 24 h and requires new synthesis of RNA and protein. A cDNA encoding ELAM-1 has been recently cloned and sequenced. ELAM-1 contains a 'lectin' domain which binds carbohydrate, a domain containing six cysteines similar to a sequence present in epidermal growth factor (EGF domain), six tandem consensus repeats, a transmembrane domain, and a short cytoplasmic tail [48]. Preliminary evidence indicates that ELAM-1 sequence is similar to a granule membrane protein 140 found in granules of platelets and EC, that is involved in platelet-EC interaction. The induction of ELAM-1 has also been documented in vivo in delayed hypersensitivity reactions [49] and in vascular leak syndrome induced by infusion of interleukin-2 in cancer patients [50].

The role of IL-1 and TNF in the process of the recruitment of leukocytes is however more complex and is not limited to the generation of signals involved in the adhesion. TNF and IL-1 also promote the synthesis of a neutrophil chemotactic factor that can provide a signal to elicit selective diapedesis of neutrophils into the extravascular space [51] and of an inhibitor of leukocyte adhesion (LAI) [16] and of an inhibitor of leukocyte chemotaxis antigenically related to the retroviral transmembrane protein p15E [52]. The production of LAI and of this inhibitor of chemotaxis represents a circuit of down-regulation in the leukocyte cytokine interaction, to limit the highly destructive changes induced by IL-1 and TNFα.

The next effects of IL-1 and TNF are those related to the modification of the balance of the coagulant-anticoagulant molecules on the EC surface toward fibrin deposition and intravascular coagulation. EC play an active role in the specific breakdown of locally deposited fibrin through the synthesis and secretion of tissue-type plasminogen activator (tPA). EC also produce type 1 plasminogen activator inhibitor (PAI-1), a fast-acting inhibitor of tPA, and thrombomodulin that participates in the anticoagulant protein C pathway. IL-1 and TNFα decrease tPA and increase PAI-1 in a dose- and time-dependent manner in EC [27–30]. An altered balance between these two activities might reduce the ability of EC to lyse intravascular thrombi and therefore limit their antithrombic properties. IL-1 and TNFα markedly increase tissue factor-like procoagulant activity in EC [17, 26, 28]. The activity is expressed on the surface of intact EC suggesting that it would be accessible to the plasma clotting system in vivo. Tissue factor activity on EC surface would lead to activation of factor VII, then of factors IX and X, propagating a procoagulant pathway on the cell surface to generate thrombin. The activation of the extrinsic pathway of coagulation is potentiated by the inhibition of the anticoagulant effect of protein C induced by both monokines [27, 28]. Precise regulation of these pathways is critical for effective hemostasis since alterations in the production of either tPA or PAI-1 have been detected during deep vein thrombosis, myocardial infarction and disseminated intravascular coagulation associated with septicemia.

Another crucial aspect is the production of autacoid substances by EC treated with IL-1 and TNFα/β. These cytokines induce the synthesis of PGI_2 and PAF (see above). PGI_2 is produced by a 'slow pathway' requiring ⩾6 h and dependent upon active synthesis of proteins [17, 23, 53], most probably cyclooxygenase. PGI_2 is a potent vasodilator and antiaggregating agent and this production may be part of a broader genetic program, as well as the modification of fibrinolytic system and adhesiveness properties of EC surface, that is coordinately activated by these two cytokines during inflammatory diseases.

EC have the capacity to produce certain cytokines, namely monocytic (M)- granulocyte (G)- and granulocyte (G)- and granulocyte-macrophage (GM)-colony-stimulating factor (CSF) [19, 20], IL-1 [18], IL-6 [54], and platelet-derived growth factor (PDGF) [21], when stimulated with IL-1 and TNFα. M-, G- and GM-CSF are glycoprotidic growth factors required for the survival, growth and differentiation of hematopoietic precursors and which affect the function of circulating mature cells. In particular, G- and GM-CSF profoundly influneces the biological activities of mature neutrophils, for example, by enhancing their ability to produce superoxide anions, to kill tumor cells, to phagocytize particles and by increasing the expression of adhesive protein (CD18 complex) [reviewed in 55]. The release of IL-1 and G-, GM-CSF by EC represents a paracrine and autocrine loop that contribute to the development of vascular diseases. Furthermore, G- and GM-CSF have specific receptors on EC surface and induce these cells to migrate, proliferate and produce tPA [31, 37]. The local production of G- and GM-CSF at sites of cell-mediated immunity and injury could be in part responsible of neovascularization processes observed in delayed hypersensitivity reactions, chronic inflammatory responses and tumor growth.

IL-1 enhances alo the basal release of IL-6 by EC associated with expression of IL-6 mRNA [54]. IL-6 is a pleiotropic lymphokine, active on T and B cells and on hepatocytes, which are stimulated by this cytokine to produce acute phase proteins [reviewed in 56]. Therefore, the ability to produce IL-6 may represent an important mechanism by which EC participate in inflammatory and immune reactions.

An important link between acute inflammatory processes and atherogenesis is the production of PDGF by TNFα-stimulated EC. TNFα releases a factor that is mitogenic for smooth muscle cells together with the expression of PDGF mRNA. This biological activity is blocked by a specific antibody anti-PDGF. Also, IL-1 controls smooth muscle cell proliferation with positive and negative effects. IL-1 stimulates the growth of these cells, but endogenous prostanoid production induced by IL-1 appears to antagonize this growth-promoting effect in short term (2 days), but not during more prolonged exposure (7–28 days) [40]. These observations indicate that inititation of atherogenesis in vivo does not require products exclusively released from platelets, and other soluble products secreted by other cells can initiate the abnormal proliferation of smooth muscle cells.

PAF/Cytokine Feedback Network

PAF belongs to a new class of lipid mediators recently identified as alkyl ether of phosphorylcholine that may be relevant in inflammation, in several hypersensitivity reactions, immunoregulation and host defense [reviewed in 6]. PAF is a potent stimulus for cells that participate in the inflammatory reactions such as platelets, neutrophils and monocytes, causing their recruitment at sites of tissue injury followed by the secretion of their granules. When injected in vivo, PAF exerts profound intravascular (thrombocytopenia, neutropenia) and systemic effects (hemoconcentration, edema, vasoconstriction, vasodilatation and increased vascular permeability).

The pattern of PAF action in the complex interplay of the immune response is not univocal as it may greatly vary depending on the type of the cellular subset that it interacts with. PAF modulates the expression of CD2 and CD3 antigens on the surface of T cells [57]. However, PAF activates suppressor cells concomitantly to an increased activity of $CD4^+$ T cells

[58]. Finally, PAF is involved in the production of IL-1 and TNFα from lipopolysaccharide-stimulated rat monocytes [59, 60]. Cytokines are known to profoundly influence cell-mediated responses, and in particular IL-1 and TNFα/β can induce the PAF synthesis in several cell types (neutrophils, monocytes, macrophages and vascular endothelium) [24, 61–66], and PAF in turn potentiates the production of these monokines from these cells [59, 60]. Furthermore, in neutrophils where PAF per se at low doses is not able to induce superoxide generation, preincubation of the cells with TNFα or GM-CSF results in a significant amplification of the biological response [67]. Given the close interplay between cytokines and PAF, a possible involvement of a pathological loop involving these mediators in tissue injury may be suggested.

PAF/Cytokine Network at Site of Microvascular Immune Injury

There is convincing in vivo evidence that PAF can directly provoke rapid neutrophil-independent increases in microvascular permeability. Intradermal injection of PAF results in a rapid localized increase in vascular permeability independent of neutrophil, mast cell or platelet activation [68]. Furthermore, evidence has been recently provided that endogenous PAF production contributes to changes in EC in postcapillary venules [69]. It is conceivable that two different mechanisms of PAF-induced increase in glomerular permeability may be operating in vivo via the release of inflammatory mediators and through its direct vasoactive properties as demonstrated by ultrastructural changes observed in glomerular EC from neutrophil- and platelet-depleted rabbits [70]. This direct action is confirmed by recent experiments performed in blood-free isolated perfused rabbit kidney showing that PAF infusion increases proteinuria independently from renal hemodynamic alterations or loss of anionic charges [71]. Also, EC damage by PAF has been observed in vivo. During PAF superinfusion of the mesenteric bed, EC reatraction occurs in the area corresponding to the site of application of PAF. As a result of this retraction, the matrix became exposed to blood and platelets adhere to these exposed areas with thrombus formation [67].

In vitro studies show that EC are a target for PAF (table 2). Recently, we have shown that PAF, but not its deacetylated and biologically inactive metabolite, lyso-PAF, dose dependently induces cultured human EC to change their shape with a modification of cytoskeleton structures. PAF-

Table 2. In vitro effects of PAF on endothelial cells

Shape change [72, 73]	– retraction
	– loss of reciprocal contact
	– elongated spindle-shaped form
Cytoskeleton impairments [72]	– disappearance of stress fibers
	– modification of vinculin distribution
Increased endothelial permeability to plasma macromolecules [72]	
Modification of calcium homeostasis [73–75]	
Translocation and activation of protein kinase C[a]	
Increased phosphoinositide turnover [75]	
Membrane hyperpolarization [76]	
Decreased adenylate cyclase activity [75]	

[a] F. Bussolino, in preparation.

treated EC retracted and tended to lose reciprocal contact [72]. Concomitant to these morphological modifications, EC grown on fibronectin-coated polycarbonate filters permit the diffusion of iodinated albumin, thus explaining, at least in part, the mechanism by which PAF increases vascular permeability in vivo [72]. These modifications of the shape and the functions are due to an increase of cytoplasmatic calcium concentration [73–76]. The rise in calcium concentration involves a production of inositol triphosphate followed by a release of the ion from intracellular stores and an influx from the extracellular compartment [73–76].

EC is not a passive target of the deleterious direct or indirect effect of PAF, as severel studies have shown that EC generate this mediator and thus can contribute to the destabilization of microenvironment. This was first demonstrated by Camussi et al. [78, 79] who showed that both hormones (angiotensin II and vasopressin) and antibodies reacting with antigens expressed on the EC membrane induce the release of PAF in vitro. Also, thrombin added repeatedly induces PAF release [78], whereas a single challenge only its synthesis [80]. Subsequent studies demonstrated that cultured EC from different sources produce PAF rapidly (few minutes) in response to leukotrienes, ATP, bradykinin, H_2O_2 and histamine [81–83].

The synthesis of PAF in EC by thrombin (and perhaps by other mentioned stimuli) involves the activation of a specific acetyltransferase that translocates a molecule of acetate from the acetyl-CoA to a molecule of lyso-PAF formed by the action of phospholipase A_2 on the alkyl-acylphosphoglycerides. The stimulus increases the V_{max} of this enzyme without affecting the K_m [84]. The activation of acetyltransferase is regulated in vivo by a phosphorylation step catalyzed by a calcium-, phospholipid-dependent protein kinase [R. Heller et al., in preparation], rather than by a cAMP-dependent protein kinase as demonstrated in vitro [85]. Activation of adenylate cyclase down-regulates acetyltransferase, thus blocking PAF production [R. Heller et al., in preparation]. After thrombin stimulation, intracellular pH increases by the activation of electroneutral Na^+/H^+ antiport that exchanges extracellular Na^+ for intracellular H^+. This alkalinization favors and regulates the increase in intracellular Ca^{2+} concentration that is essential for in vivo activation of the acetyltransferase.

In contrast to the above mentioned stimuli that cause a virtually immediate production of PAF, we have recently elucidated a new pathway that requires protein synthesis and is primed by IL-1 and TNFα/β [24, 61–63]. In this pathway, PAF is produced after 2–8 h after addition of the cytokines and about 20–30% is released in the medium (fig. 1). The kinetics of PAF synthesis is faster and more PAF is produced in TNFα/β-treated cells than in those challenged with IL-1. However, PAF synthesis is stimulated by a lower concentration of IL-1 than TNFα.

This 'slow' pathway for PAF synthesis has also been described in monocytes stimulated with immune interferon, IL-1 and TNFα [65], whereas in neutrophil and in macrophages the time course of the PAF production after IL-1 and TNFα stimulation is rapid [63]. Also, these cytokines activate the specific acetyltransferase, but its possible mechanism of action is unknown. Recently, we demonstrated that chloromethyl ketone proteinase inhibitors and plasma α_1-proteinase inhibitor or α_1-antichymotrypsin inhibit the PAF production induced by IL-1 and TNFα in neutrophils, EC and macrophages and that proteinases mimic the activation by cytokines [64, 66]. A hypothesis consistent with our findings is that IL-1 and TNFα activate a protease that cleaves lipocortins or related molecules on the cell surface and may promote the activation of phospholipase A_2 leading the production of lyso-PAF. Alternatively, these proteinases may inactivate an inhibitor of the acetyltransferase. These experiments also show that PAF production may be inhibited by acute phase inhibitor (α_1-proteinase inhibitor and α_1-antichymotrypsin) at concentrations 10- to 100-fold lower than those present in normal serum. Therefore, it is unlikely that neutrophils, macrophages and EC produce PAF when exposed to the vast excess of antiproteinase present in the serum. Synthesis of PAF requires the inactivation of these proteinase inhibitors. The treat-

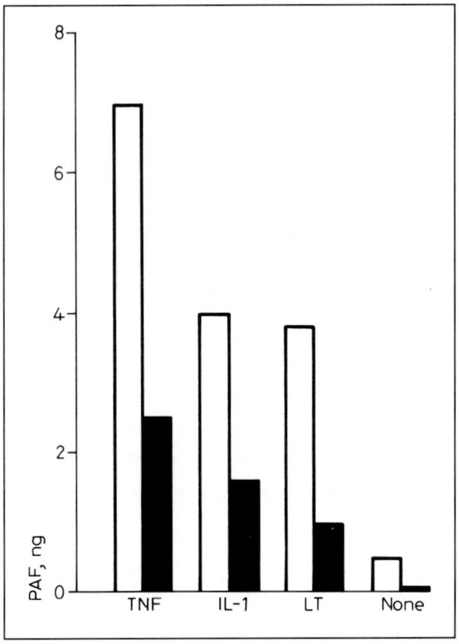

Fig. 1. PAF production by EC (5×10^5 cells) stimulated with 0.2 n*M* IL-1, TNF and LT for 6 h at 37 °C. PAF was extracted, characterized and measured as described before [62]. Solid bar: PAF released; open bar: PAF associated with EC. This figure represents a typical experiment.

ment of these proteins with oxidant prevents their ability to block the PAF production, suggesting that O_2 radicals produced at the site of inflammation can inactivate these antiproteinases permitting the synthesis of PAF. In view of these findings, it seems possible that PAF is only produced in zones of acute inflammation or where the cell plasma membranes adhere to other cells or to the extracellular matrix, thus excluding plasma proteins.

In in vivo experimental conditions characterized by an increased PAF production, the biosynthesis of this mediator is closely linked to the increased activity of biosynthetic enzymes [reviewed in 6]. In vivo the enhanced synthesis may be balanced by serum antiproteinases that prevent the enhanced activity of biosynthetic enzymes [64, 66]. Once formed, however, PAF may be inactivated intracellularly or in plasma by the specific PAF acetylhydrolase that cleaves the acetate moiety from the PAF molecule giving the biological inactive derivative lyso-PAF [87, 88]. Therefore, the biological relevance of PAF may be related to an increased synthesis due to appropriate stimulation and/or inactivation of serum inhibitors of PAF biosynthesis or to reduced degradation. Due to the presence of these multiple regulatory systems, the detectability of PAF in plasma is limited to acute pathological conditions such as anaphylaxis [89], primary acquired cold urticaria [90], allograft rejection [91], sepsis [92], and acute serum sickness [93].

For the above mentioned reasons, the biological significance of PAF produced and released by EC and other cell types might be limited to the microvascular circulation. PAF released by EC can activate circulating cells (platelets, neutrophils, monocytes), or spread toward the smooth muscle cells eliciting their activation [94], or act on endothelium itself with an autocrine mechanism. PAF produced, but not released, by EC is partially located in the outer membrane, thus mediating the adhesion of neutrophils [80–83]. Furthermore, the adhesion of neutrophils to EC stimulated by IL-1 is characterized by a production of PAF by neutrophils themselves other than by EC. It is possible that EC stimulated by IL-1 express molecules activating neutrophils to produce PAF [95].

A limited and recent number of studies indicate that cytokines and PAF are involved in several pathological conditions. Sun and Hsueh [96] demonstrated that PAF receptor antagonists can prevent TNFα-induced bowel necrosis in rat and that TNFα induces PAF synthesis in this tissue. In a modified model of acute serum sickness in the rabbit in which phytohemagglutinin or TNFα or IL-1 are injected in the early phase of immune complex deposition, Camussi et al. [in preparation] observed a massive accumulation of monoculear cells in glomeruli, resulting in a more persistent proteinuria associated with a significant decrease of renal function. Isolated glomeruli produce higher amounts of TNF and IL-1 than glomeruli isolated from rabbits having acute serum sickness. The enhanced production of monokines correlates with the progression of glomerular disease and with the increased glomerular levels of PAF-specific acetyltransferase. PAF receptor antagonists partially block the glomerular injury. TNFα and PAF have been demonstrated to be mediators involved in septic shock [92, 97]. PAF receptor antagonists [6] and antibody anti-TNFα [97] attenuate experimental septic shock. Huer and Letts [98] have shown that pretreatment of mice with endotoxin or TNFα enhanced the mortality induced by PAF. Recently we have focused our attention on diabetic retinopathy and on keratouveitis, two ocular diseases characterized by an

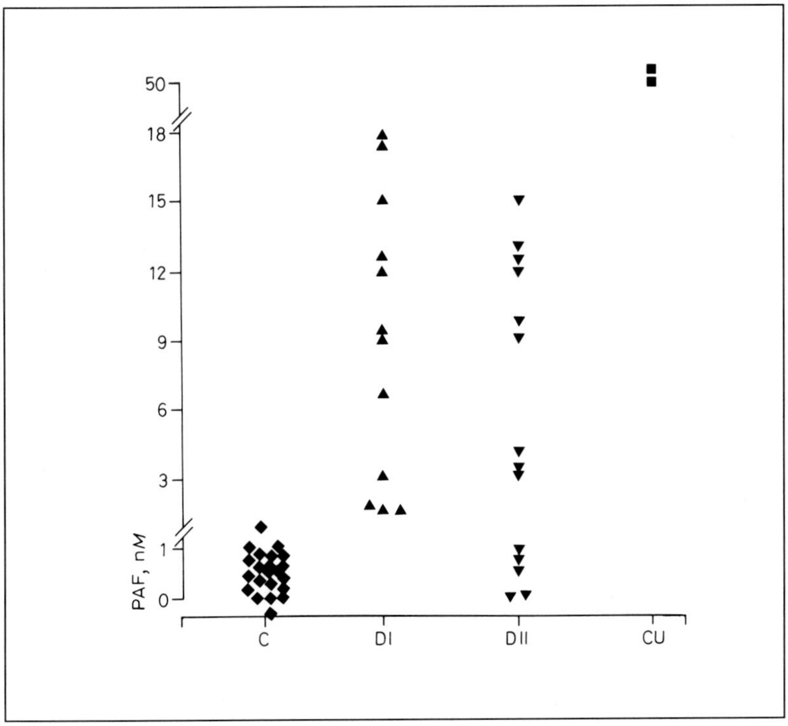

Fig. 2. PAF concentration in *humor aqueous.* C = Control; DI = type I diabetes mellitus; DII = type II diabetes mellitus; CU = keratouveitis.

altered immune response and a modification of 'polypeptide network' [99, 100]. We observed a high level of PAF in the *humor aqueous* of these patients (fig. 2). In diabetic subjects the PAF concentration is 5- to 10-fold higher than the control and is unrelated to the glycemia. Preliminary observations indicate a relationship between the high level of PAF and the severity of microangiopathy, suggesting that PAF/cytokine networks could play a role also in chronic microvascular diseases. This hypothesis is supported by results showing that IL-1, TNFα and PAF injected intravitreally have inflammatory effects characterized by edema, hemorrhage, cellular infiltration, vasopermeabilization and neovascularization [101–103]. Furthermore, a PAF receptor antagonist prevents the inflammatory effects of IL-1 in the rabbit eye [104]. The complex circuits that link PAF and cytokines are summarized in table 3.

Table 3. Interactions between PAF and cytokine networks

IL-1 and TNF-α/β induce the PAF synthesis by EC, neutrophils monocytes and macrophages
PAF modulates the synthesis of IL-1 and TNF-α by monocytes
TNF and IL-6 increase the production of proteins of acute phase, namely α_1-proteinase inhibitor and α_1-antichymotrypsin, that inhibit the PAF production induced by TNF and IL-1
IL-1 and PAF can modulate the immune response by acting directly on T cells
The activation of several cell types is triggered by PAF, IL-1, TNF-α, G- and GM-CSF
TNFα and GM-CSF prime neutrophil response induced by PAF
TNFα and PAF are mediators of sepsis and PAF receptor antagonists, and antibody anti-TNF prevent septic shock
GM-CSF primes neutrophils to produce more PAF after stimulation with chemoattractants

Conclusion

PAF and cytokines may be important primary initiators of microvascular injury, which may be operative in a diverse range of pathological processes. The use of new therapeutic approaches that utilize PAF receptor antagonists [reviewed in 6] or modulate PAF production or synthetic peptides that block the effects of cytokines by interacting with their cellular receptors, should be useful in the management of microvascular diseases.

References

1 Ross R: The pathogenesis of atherosclerosis: An update. N Engl J Med 1986;314: 488–496.
2 Munro JM, Cotran RS: Biology of disease. The pathogenesis of atherosclerosis: atherogenesis and inflammation. Lab Invest 1988;58:249–261.
3 Sporn MB, Roberts AB: Peptide growth factors and inflammation, tissue, repair, and cancer. J Clin Invest 1986;78:329–332.
4 Rodgers GM: Hemostatic properties of normal and perturbed vascular cells. FASEB J 1988;2:116–123.
5 Fauci AS, Haynes BF, Katz P: The spectrum of vasculitis: clinical, pathologic, immunologic and therapeutic considerations. Ann Intern Med 1978;89:660–688.
6 Braquet P, Touqui L, Shen TS, et al: Perspectives in platelet-activating factor research. Pharmacol Rev 1987;39:97–145.

7 Baldwin WM: The symbiosis of immunocompetent and endothelial cells. Immunol Today 1982;3:267–270.

8 Collins T, Korman AJ, Wake CT, et al: Immune interferon activates multiple class II major histocompatibiltiy complex antigens and the associated invariant chain gene in human endothelial cells and dermal fibroblasts. Proc Natl Acad Sci USA 1984;81: 4917–4921.

9 Lapierre LA, Fiers W, Pober JS: Three distinct classes of regulatory cytokines control endothelial cell MHC antigen expression. Interaction with immune interferon differentiates the effects of tumor necrosis factor and lymphotoxin from those of leukocyte and fibroblast interferons. J Exp Med 1988;167:794–804.

10 Stolpen AH, Guinan EC, Fiers W, et al: Recombinant tumor necrosis factor and immune interferon act singly and in combination to reorganize human vascular endothelial cell monolayers. Am J Pathol 1986;123:16–24.

11 Montesano R, Orci L, Vassalli P: Human endothelial cell culture: phenotypic modulation by leukocyte interleukins. J Cell Physiol 1984;122:424–434.

12 Cavender DE, Haskard DO, Joseph B, et al: Interleukin 1 increases the binding of human B and T lymphocytes to endothelial cell monolayer. J Immunol 1986;136: 203–207.

13 Bevilacqua MP, Pober JS, Mendrick DL, et al: Identification of an inducible endothelial-leukocyte adhesion molecule. Proc Natl Acad Sci USA 1987;84:9238–9242.

14 Pober JS, Gimbrone MA, Lapierre LA, et al: Overlapping pattern of activation of human endothelial cells by interleukin 1, tumor necrosis factor and immune interferon. J Immunol 1986;137:1893–1896.

15 Gamble JR, Vadas MA: Endothelial adhesiveness for blood neutrophils is inhibited by transforming growth factor-β. Science 1988;242:97–99.

16 Wheeler ME, Luscinskas FW, Bevilacqua MP, et al: Cultured human endothelial cells stimulated with cytokines or endotoxin produce an inhibitor of leukocyte adhesion. J Clin Invest 1988;82:1211–1218.

17 Dejana E, Breviario F, Erroi A, et al: Modulation of endothelial cell function by different molecular species of interleukin 1. Blood 1987;69:695–699.

18 Libby P, Ordovas JM, Auger KR, et al: Endotoxin and tumor necrosis factor induce IL-1 gene expression in adult human vascular endothelial cells. Am J Pathol 1986; 124:179–185.

19 Zsebo KM, Yushenkoff VN, Schiffer S, et al: Vascular endothelial cells and granulopoiesis: interleukin-1 stimulates release of G-CSF and GM-CSF. Blood 1988;71: 99–103.

20 Broudy V, Harlam JM, Adamson JW: Disparate effects of tumor necrosis factor-/cachectin and tumor necrosis factor-/lymphotoxin on hematopoietic growth factor production and neutrophil adhesion molecule expression by cultured human endothelial cells. J Immunol 1987;138:4298–4302.

21 Hajjar KA, Hajjar DP, Silverstein RL, et al: Tumor necrosis factor-mediated release of platelet-derived growth factor from cultured endothelial cells. J Exp Med 1987; 166:235–245.

22 Ripoche J, Mitchell JA, Madin EC, et al: Interferon-induced synthesis of complement alternative pathway proteins by human endothelial cells in culture. J Exp Med 1988;168:1917–1922.

23 Kawakami M, Ishibashi S, Ogawa H, et al: Cachectin/TNF as well as interleukin-1 induces prostacyclin synthesis in cultured vascular endothelial cells. Biochem Biophys Res Commun 1986;141:482–488.
24 Bussolino F, Camussi G, Tetta G, et al: Selected cytokines promote the synthesis of platelet-activating factor in vascular endothelial cells: comparison between tumor necrosis factor and interleukin 1. J Lipid Mediat, in press.
25 Frasier-Scott K, Hatzakis H, Seong D, et al: Influence of natural and recombinant interleukin 2 on endothelial cell arachidonate metabolism. Induction of de novo synthesis of prostaglandin H synthase. J Clin Invest 1988;82:1877–1883.
26 Bevilacqua MP, Pober JS, Majeau GR, et al: Recombinant tumor necrosis factor induces procoagulant activity in cultured human endothelial cells: characterization and comparison with the actions of interleukin 1. Proc Natl Acad Sci USA 1986;83: 4533–4537.
27 Naworth PP, Stern DM: Modulation of endothelial cell hemostatic properties by tumor necrosis factor. J Exp Med 1986;163:740–745.
28 Naworth PP, Handley CT, Esmon C, et al: Interleukin 1 induced endothelial cell procoagulant activity while suppressing cell surface anticoagulant activity. Proc Natl Acad Sci USA 1986;83:3460–3464.
29 Emeis JJ, Kooistra T: Interleukin 1 and lipopolysaccharide induce an inhibitor of tissue-type plasminogen activator in vivo and in cultured endothelial cells. J Exp Med 1986;163:1260–1265.
30 Schleef RS, Bevilacqua MP, Sawdey M, et al: Cytokine activation of vascular endothelium. Effects on tissue-type plasminogen activator and type 1 plasminogen activator inhibitor. J Biol Chem 1988;5797–5803.
31 Kojima S, Tadenuma H, Inada Y, et al: Enhancement of plasminogen activator activity in cultured endothelial cells by granulocyte colony-stimulating factor. J Cell Physiol 1989:138:192–196.
32 Frater-Schroder M, Risau W, Hallmann R, et al: Tumor necrosis factor type α, a potent inhibitor of endothelial cell growth in vitro, is angiogenic in vivo. Proc Natl Acad Sci USA 1987;84:5277–5281.
33 Norioka K, Hara M, Kiani A, et al: Inhibitory effect of human recombinant interleukin-1 and on growth of human vascular endothelial cells. Biochem Biophys Res Commun 1987;145:969–975.
34 Takehara K, LeRoy EC, Grotendorst GR: TFG-β inhibition of endothelial cell proliferation: alteration of EFG binding and EGF-induced growth-regulatory (competence) gene expression. Cell 1987;49:415–522.
35 Schreiber AB, Winkler ME, Derynck R: Transforming growth factor-alpha: a more potent angiogenic mediator than epidermal growth factor. Science 1986;232:1250–1253.
36 Folkman J, Klagsbrun M: Angiogenic factors. Science 1987;236:442–447.
37 Bussolino F, Wang JM, Defilippi P, et al: Granulocyte- and granulocyte-macrophage colony-stimulating factors induce human endothelial cells to migrate and proliferate. Nature 1989;337:471–473.
38 Montesano R, Mossaz A, Ryser JE, et al: Leukocyte interleukins induce cultured endothelial cells to produce a highly organized, glycosaminoglycan-rich pericellular matrix. J Cell Biol 1984;99:1706–1715.
39 Ross R, Rainers EW, Bowen-Pope DF: The biology of platelet-derived growth factor. Cell 1986;46:155–169.

40 Libby P, Warner SJC, Friedman GB: Interleukin 1: a mitogen for human vascular smooth muscle cells that induces the releases of growth-inhibitory prostanoids. J Clin Invest 1988;81:487–498.

41 Hansson G: Vascular effects of cytokines; in European Artery Club: 10th Anniversary Conference on Vascular Biology, Cambridge 1989.

42 Assoian RK, Sporn MB: Transforming growth factor-beta acts on aortic smooth muscle cells; in Fidge NH, Nestel PJ (eds): Atherosclerosis, VII. Amsterdam, Excerpta Medica, 1986, p 459.

43 Berk BC, Alexander RW, Brock TA, et al: Vasoconstriction: a new activity for platelet-derived growth factor. Science 1986;232:87–90.

44 Warener SJC, Libby P: Human vascular smooth muscle cells. Target for and source of tumor necrosis factor. J Immunol 1989;142:100–109.

45 Harlam JM: Leukocyte-endothelial cell interactions. Blood 1985;65:513–525.

46 Stoolman LM: Adhesion molecules controlling lymphocyte migration. Cell 1989;56: 907–910.

47 Gamble JR, Harlam JM, Klebanoff SJ, et al: Stimulation of the adherence of neutrophils to umbilical vein endothelium by recombinant tumor necrosis factor. Proc Natl Acad Sci USA 1985;82:8667–8671.

48 Bevilacqua MP, Stengelin S, Gimbrone MA: Endothelial leukocyte adhesion molecule 1: an inducible receptor for neutrophils related to complement regulatory proteins and lectins. Science 1989;243:1160–1164.

49 Cotran RS, Gimbrone MA, Bevilacqua MP, et al: Induction and detection of a human endothelial activation antigen in vivo. J Exp Med 1986;164:661–666.

50 Cotran RS, Pober JS, Gimbrone MA, et al: Endothelial activation during interleukin 2 (IL-2) immunotherapy: a possible mechanism for the vascular leak syndrome. J Immunol 1988;139:1883–1888.

51 Strieter RM, Kunkel SL, Showell HJ, et al: Endothelial cell gene expression of a neutrophil chemotactic factor by TNF-α, LPS, and IL-1. Science 1989;243:1467–1469.

52 Wang JM, Chen ZC, Cianciolo GJ, et al: Production of a retroviral P15E-related chemotaxis inhibitor by IL-1-treated endothelial cells: a possible negative feedback in the regulation of the vascular response to monokines. J Immunol 1989;142:2012–2017.

53 Rossi V, Breviario F, Ghezzi P, et al: Prostacyclin synthesis induced in vascular cells by interleukin-1. Science 1985;229:174–177.

54 Sironi M, Breviario F, Proserpio A, et al: IL-1 stimulates IL-6 production in endothelial cells. J Immunol 1989;142:549–553.

55 Sieff CA: Hematopoietic growth factors. J Clin Invest 1987;79:1549–1557.

56 Wong GC, Clark SC: Multiple actions of interleukin 6 within a cytokine network. Immunol Today 1988;9:137–139.

57 Vivier E, Salem P, Dulioust A, et al: Immunoregulatory functions of PAF-acether. II. Decrease of CD2 and CD3 antigen expression. Eur J Immunol 1988;18:425–430.

58 Rola-Pleszczynski M, Pouliot C, Turcotte S, et al: Immune regulation by platelet-activating factor. I. Induction of suppressor cell activity in human monocytes and $CD8^+$ cells and helper activity in $CD4^+$ T cells. J Immunol 1988;140:3547–3552.

59 Ward SG, Lewis PG, Westwick J: Platelet-activating factor stimulates interleukin 1 production from human adherent monocyte-macrophages; in Braquet P (ed): New Trends in Lipid Mediators Research. Platelet-Activating Factor and Cell Immunology (Basel, Karger, 1988, pp 6–15).
60 Bonavida B, Mencia-Huerta JM, Braquet P: Effect of platelet-activating factor (PAF) on monocyte activation and production of tumor necrosis factor (TNF). J Allergy Appl Immunol 1989;88:157–160.
61 Bussolino F, Breviario F, Tetta C, et al: Interleukin 1 stimulates platelet-activating factor production in cultured human endothelial cells. J Clin Invest 1986;77:2027–2033.
62 Bussolino F, Camussi G, Baglioni C: Synthesis and release of platelet-activating factor by human vascular endothelial cells treated with tumor necrosis factor or interleukin 1. J Biol Chem 1988;263:11856–11861.
63 Camussi G, Bussolino F, Salvidio G, et al: Tumor necrosis factor/cachectin stimulates peritoneal macrophages, polymorphonuclear neutrophils and vascular endothelial cells to synthesize and release platelet-activating factor. J Exp Med 1987;166: 1390–1404.
64 Camussi G, Tetta C, Bussolino F, et al: Synthesis and release of platelet-activating factor is inhibited by plasma α_1-proteinase inhibitor or α_1-antichymotrypsin and is stimulated by proteinases. J Exp Med 1988;168:1293–1306.
65 Valone FH, Epstein LB: Biphasic platelet-activating factor synthesis by human monocytes stimulated with IL-1, tumor necrosis factor, or IFN-γ. J Immunol 1988; 140:3945–3950.
66 Camussi G, Tetta C, Bussolino F, et al: Tumor necrosis factor stimulates human neutrophils to release leukotriene B_4 and platelet-activating factor. Induction of phospholipase A_2 and acetyl-CoA: 1-alkyl-sn-glycero-3-phosphocholine O_2-acetyltransferase activity and inhibition by antiproteinases. Eur J Biochem 1989;182:661–666.
67 Braquet P, Paubert Braquet M, Bourgain R, et al: Hypothesis. PAF/cytokine autogenerated feedback networks in microvascular immune injury: consequences in shock, ischemia and graft rejection. J Lipid Mediat 1989;1:75–112.
68 Humprey DM, McManus L, Hanahan DJ, et al: Morphologic basis of increased vascular permeability induced by acetyl glyceryl ether phosphorylcholine. Lab Invest 1984;50:16–25.
69 Hellewell PG, Williams TJ: A specific antagonist of platelet-activating factor suppresses oedema formation in an Arthus reaction but no oedema induced by leukocyte chemoattractants in rabbit skin. J Immunol 1986;137:302–307.
70 Camussi G: Potential role of platelet-activating factor in renal pathophysiology. Kidney Int 1986;29:469–477.
71 Perico N, Delaini F, Tagliaferri M, et al: Effect of platelet-activating factor (PAF) and its specific receptor antagonists on glomerular permeability in isolated perfused rat kidney. Lab Invest 1988;58:163–171.
72 Bussolino F, Camussi G, Aglietta M, et al: Human endothelial cells are target for platelet-activating factor. I. Platelet-activating factor induces changes in cytoskeleton structures. J Immunol 1987;139:2439–2446.
73 Grigorian GY, Ryan US: Platelet-activating factor effects on bovine pulmonary artery endothelial cells. Circ Res 1987;61:389–395.

74 Bussolino F, Aglietta M, Sanavio F, et al: Alkyl-ether phosphoglycerides influence calcium fluxes into human endothelial cells. J Immunol 1985;135:2748–2755.
75 Brock TA, Gimbrone MA: Platelet-activating factor alters calcium homeostasis in cultured vascular endothelial cells. Am J Physiol 1986;250:H1086–H1092.
76 Ryan US, Grigorian PV, Avdonin PV, et al: Platelet-activating factor effects on endothelial cells; in Braquet P (ed): New Trends in Lipid Mediators Research. Platelet-Activating Factor and Cell Immunology, Karger, Basel, 1988, pp 144–158.
77 Lerner R, Lindstrom P, Palmblad J: Platelet-activating factor and leukotriene B_4 induce hyperpolarization of human endothelial cells but depolarization of neutrophils. Biochem Biophys Res Commun 1988153:805–810.
78 Camussi G, Aglietta M, Malavasi F, et al: The release of platelet-activating factor from human endothelial cells in culture. J Immunol 1983;131:2397–2403.
79 Camussi G, Pawloski I, Bussolino F, et al: Release of platelet-activating factor in rabbits with antibody-mediated injury of the lung. The role of polymorphonuclear neutrophils and pulmonary endothelial cells. J Immunol 1983;131:1802–1807.
80 Prescott SM, Zimmerman GA, McIntyre TM: Human endothelial cells in culture produce platelet-activating factor (1-alkyl-2-acetyl-sn-glyceryl-3-phosphocholine) when stimulated with thrombin. Proc Natl Acad Sci USA 1984;81:3934–3938.
81 McIntyre TM, Zimmerman GA, Setoh K, et al: Cultured endothelial cells synthesize both platelet-activating factor and prostacyclin in resonse to histamine, bradykinin and adenosine triphosphate. J Clin Invest 1985;76:271–280.
82 McIntyre TM, Zimmerman GA, Prescott SM: Leukotriene C_4 and D_4 stimulate human endothelial cells to synthesize platelet-activating factor and bind neutrophils. Proc Natl Acad Sci USA 1986;83:2204–2206.
83 Lewis MR, Whatley RE, Cain P, et al: Hydrogene peroxide stimulates the synthesis of platelet-activating factor by endothelial cells and induces endothelial cell-dependent neutrophil adhesion. J Clin Invest 1988;82:1220–1228.
84 Hirafuji M, Mencia-Huerta JM, Benveniste J: Regulation of PAF-acether (platelet-activating factor) biosynthesis in cultured human endothelial cells stimulated with thrombin. Biochim Biophys Acta 1987;930:359–369.
85 Gomez-Cambronero J, Velasco S, Mato JM, et al: Modulation of lyso-platelet-activating factor:acetyl-CoA acetyltransferase from rat splenic microsomes. The role of cyclic AMP-dependent protein kinase. Biochim Biophys Acta 1985;845:516–519.
86 Ghigo D, Bussolino F, Garbarino G, et al: Role of Na^+/H^+ exchange in thrombin-induced platelet-activating factor production by human endothelial cells. J Biol Chem 1988;263:19437–19446.
87 Blank ML, Lee T-C, Fitzgerald V, et al: A specific acetylhydrolase for 1-alkyl-2-acetyl-sn-glycero-3-phosphocholine (a hypotensive and platelet-activating lipid). J Biol Chem 1981;256:175–178.
88 Farr RS, Wardlow ML, Cox CP, et al: Human serum acid-labile factor is an acylhydrolase that inactivates platelet-activating factor. Fed Proc 1983;42:3120–3122.
89 Pinckard RN, Farr RS, Hanahan DJ: Physicochemical and functional identity of rabbit platelet-activating factor (PAF) released in vivo during IgE anaphylaxis with PAF released in vitro from IgE-sensitized basophils. J Immunol 1979;123:1847–1857.
90 Grandel KE, Farr RS, Wanderer AA, et al: Association of platelet-activating factor with primary acquired cold urticaria. N Engl J Med 1985;313:405–408.

91 Ito S, Camussi G, Tetta C, et al: Hyperacute renal allograft rejection in the rabbit. The role of platelet-activating factor and of cationic proteins derived from poylmorphonuclear leukocytes and from platelets. Lab Invest 1984;51:148–161.
92 Bussolino F, Porcellini MG, Varese L, et al: Intravascular release of platelet-activating factor in children with spesis. Thromb Res 1987;48:619–620.
93 Camussi G, Tetta C, Deregibus C, et al: Platelet-activating factor (PAF) in experimentally-induced rabbit acute serum sickness: role of basophil-derived PAF in immune complex deposition. J Immunol 1982;128:86–94.
94 Schwertschlag US, Whorton AR: Platelet-activating factor-induced homologous and heterologous desensitization in cultured vascular smooth muscle cells. J Biol Chem 1988;263:12791–13796.
95 Breviario F, Bertocchi F, Dejana E, et al: Interleukin-1-induced adhesion of polymorphonuclear leukocytes to cultured human endothelial cells. Role of platelet-activating factor. J Immunol 1988;141:3391–3397.
96 Sun X, Hsueh W: Bowel necrosis induced by tumor necrosis factor in rats is mediated by platelet-activating factor. J Clin Invest 1988;81:1328–1331.
97 Cerami A, Beutler B: The role of cachectin/TNF in endotoxic shock and cachexia. Immunol Today 1988;9:28–31.
98 Heuer H, Letts G: Priming effects of PAF in vivo by tumor necrosis factor and endotoxin. J Lipid Mediat, in press.
99 Eisenbarth GS: Type I diabetes mellitus: a chronic autoimmune disease. N Engl J Med 1986;314:1360–1368.
100 Hooks JJ, Chan CC, Detrick B: Identification of the lymphokines, interferon-gamma and interleukin-2 in inflammatory eye diseases. Invest Ophthalmol Vis Res 1988;29:1444–1451.
101 Rosenbaum JT, Samples HR, Hefeneider SH, et al: Ocular inflammatory effects of intravitreal interleukin 1. Arch Ophthalmol 1987;105:1117–1120.
102 Rosenbaum JT, Howes EL, Rubin RM, et al: Ocular inflammatory effects of intravitreally injected tumro necrosis factor. Am J Pathol 1988;133:47–53.
103 Braquet P, Vidal RF, Braquet M, et al: Involvement of leukotrienes and PAF-acether in the increased microvascular permeability of the rabbit retina. Agents Actions 1984;19:82–85.
104 Rubin RM, Rosenbaum JT: A platelet-activating factor antagonist inhibits interleukin 1-induced inflammation. Biochim Biophys Res Commun 1988;154:429–436.
105 Wirthmuller U, De Weck AL, Dehinden CA: Platelet activating factor production in human neutrophils by sequential stimulation with granulocyte-macrophage colony stimulating factor and the chemotactic factors C5α or formyl-methionyl-leucyl-phenylalanine. J Immunol 1989;142:3213–3218.

Dr. F. Bussolino, Dipartimento di Genetica, Biologica e Chimica Medica,
V. Santena 5bis, I–10126 Torino (Italy)

Bazan NG (ed): Lipid Mediators in Eye Inflammation.
New Trends Lipid Mediators Res. Basel, Karger, 1990, vol 5, pp 149–167

Electroretinogram and Mediators of Inflammation

Michel Doly[a], *Ginette Gaillard*[a], *Pierre Braquet*[b]

[a] Laboratoire de Biophysique, Facultés de Médecine et de Pharmacie, Unité INSERM U.71, Clermont-Ferrand, and [b] Institut Henri Beaufour (IHB), Research Laboratories, Le Plessis Robinson, France

Ocular immunology now appears to be a preferential field of ophthalmological research which may allow us to understand better the mechanisms of numerous pathological processes. In these conditions, it is important to specify the effects of the known immunomediators on visual perception. For this purpose, the electrophysiological approach seems to be one of the best adapted. Indeed, the electroretinogram (ERG) constitutes the objective and functional response of the retina to the light, its natural stimulus. So, this receptor potential represents a quantitative signal which directly reflects the quality of visual function. In these coditions, every molecule able to modify ERG may interact with visual perception. In addition, every inflammatory mediator present inside the retina or which can be released at its level, and acting upon ERG, may be a messenger between ocular inflammatory impairments and visual function. Therefore, every molecule able to interfere in this specific relation between inflammation and visual perception will be able to reduce consequential effects of any inflammatory process on retinal function.

This is the reason why we experimented with several inflammatory mediators in order to establish their effects on the ERG. Willing to work with a simple signal, we chose to experiment on isolated retina. Doing so, we can record the response of retinal to the light stimulation without the participation of any other ocular tissues. Furthermore, this experimental set-up allows us to put the tested molecule directly in presence of the retina. In order to work with experimental conditions comparable to human ones, we used isolated mammalian retina. As we shall see, the results obtained will be interpreted only in accordance with the molecular mechanisms which generate ERG. Nowadays, such mechanisms are well known in the field of scotopic vision, which allows us to appreciate various effects of inflammatory mediators at several levels of ERG generation.

In a first phase we experimented with hydroxyl radical ($OH^{\bullet}$) [1] by inducing radical production by means of the nonenzymatic catalytic system (Fe^{2+} + ascorbate) proposed by Kagan et al. [2]. As a matter of fact, it is well established that the retinal membranes are particularly rich in polyunsaturated fatty acids and therefore are especially sensitive to oxygenated free radicals [3]. So, the retina constitutes an excellent model to study the induction of membrane peroxidation and its inhibition by radical scavengers such as *Ginkgo biloba* extract that we have tested.

In a second phase we experimented with leukotrienes B_4, C_4 and D_4. These molecules are biologically active arachidonate metabolites discovered in polymorphonuclear leukocytes by Borgeat and Samuelsson [4]. The key enzyme of this pathway is a 5-lipoxygenase which catalyzes the transformation of arachidonic acid to 5-hydroperoxyeicosatetraenoic acid. Subsequent metabolism leads to the formation of leukotrienes A_4, B_4, C_4, D_4 and E_4. These compounds are released by various tissues and cellular preparations following immunological and chemical stimulations: those are believed to play a fundamental role in inflammatory and allergic reactions. Among thier many biological effects, leukotrienes modify the membrane ionic exchanges both at the cellular level and in artificial membrane models [5, 6]; this is the reason why we experimented them on the retinal function.

Finally, we performed numerous experiments with platelet-activating factor ((R)PAF, PAF-acether, PAF, AGEPC). It is a well-known phospholipid mediator identified as 1-O-alkyl-2-(R)-acetylglycero-3-phosphorylcholine [7]. PAF-acether is released by many tissues following immunological stimulation and plays a key role in inflammation and anaphylaxis [for review, see 8]. PAF is able to modify membrane ionic exchange as demonstrated by: (1) impairment of potassium transport in human red cells and mouse macrophages [9]; (2) induction of smooth muscle cell contraction, and (3) disturbances of heart contractility [reviewed in 8].

Material and Methods

Description of the Experimentation with Isolated Retina

Retinas of albino rats were used since this animal has only a scotopic visual system constituted by a single type of photoreceptor, the rods, making the ERG interpretation easier [10].

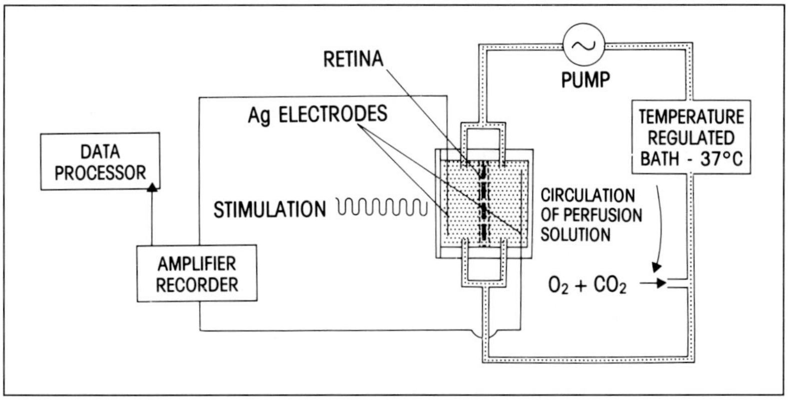

Fig. 1. Experimental set-up to record electrophysiological response of the isolated rat retina.

The method used for maintaining the function of isolated albino rat retinas is a modified version of the one used by Arden and Ernst [11] for pigeon retina. The rats were dark adapted for at least 2 h and then sacrificed. One eye was enucleated and then incised along its equator. In the perfusion liquid the retina was progressively detached from the retinal epithelium and finally freed by cutting the optic nerve. The retina was then mounted as a membrane in a perfusion cell (see figure 1 for details, and Doly et al. [12] for description). The perfusion liquid had the following composition: NaCl, 121.5 m*M*; KCl, 3.1 m*M*; $CaCl_2$, 0.4 m*M*; $MgSO_4$, 0.6 m*M*; $NaHCO_3$, 24.4 m*M*; NaH_2PO_4, 0.5 m*M*; sodium glutamate, 0.5 m*M*; glucose, 28 m*M*, freshly reconstituted human plasma 3%. A mixture consisting of 95% O_2 and 5% CO_2 was continuously bubbled in the solution, which was maintained at 37 °C.

The variation of the transretinal potential (ERG) was recorded by means of an amplifier-recorder system. The retina was stimulated with white light (300 lx, 1 ms) directly falling onto the rod outer segments.

Experiments with Hydroxyl Radicals (OH˙)

To induce formation of hydroxyl radicals and retinal membranes lipoperoxidation, a mixture of (10 ml Fe SO_4, 6.25 m*M* + 10 ml ascorbate Na, 0.25 *M*) was added to the total volume (1 liter) of the perfusion liquid. In these conditions, the radical scavenger property of *G. biloba* extract was tested by addition, into the perfusion solution, of both the peroxidative system and 10 ml of aqueous solution of extract (10 g/l).

Experiments with Leukotrienes

To investigate the possible effects of leukotrienes on ERG we have experimented with leukotriene B_4 (LTB_4), C_4 (LTC_4) and D_4 (LTD_4) (a generous gift from Dr. J. Rokash, Merck Frosst, Montreal, Canada). These molecules were added to the perfusion medium at the following concentrations: LTB_4: 2×10^{-8} *M*; 10^{-7} *M*; 4×10^{-7} *M*; LTC_4 and LTD_4: 2×10^{-8} *M*; 6×10^{-8} *M*; 10^{-7} *M*.

Experiments with PAF-Acether

In order to assess the specificity of (R)PAF-acether effects on the retina, a specific antagonist (Ginkgolide B, BN 52021) [13], (S)PAF-acether, the ineffective enantiomer, and lyso-PAF were used. (R)PAF-acether (1-O-hexadecyl-2-(R)-acetylglycero-3-phosphorylcholine), and lyso-PAF were purchased from Bachem Feinchemikalien AG (Switzerland). (S)PAF-acether, the inactive enantiomer, was a generous gift of Prof. J.J. Godfroid (Paris). They were solubilized into the perfusion medium and added directly to the perfusion solution in a volume of approximately 1 ml (for 1 liter of solution). The concentrations tested were 2×10^{-11} *M*; 2×10^{-9} *M* and 2×10^{-7} *M* for (R)PAF-acether and 2×10^{-7} *M* for (S)PAF-acether and lyso-PAF. Ginkgolide B (BN 52021: 9H-1, 7a-(epoxymethano)-1H,6aH, cyclopenta[c]furo[2,3-b]furo-[3′,2′:3,4]cyclopenta[1,2-d]furan-5,9,12-[4H]-trione, 3-tert-butylhexahydro-4,7b,-11-trihydroxy-8-methyl) was supplied by Institut Henri Beaufour (Le Plessis Robinson, France). It was solubilized in the perfusion medium and added to the solution for experiments in a volume of about 1 ml. The concentration tested was 2×10^{-5} *M* for BN 52021 in presence of 2×10^{-7} *M* (R)PAF-acether.

The interpretation of results obtained with PAF-acether on isolated retina has suggested the existence of specific PAF-binding sites in this tissue which have been characterized as follows:

(1) *Preparation of membrane homogenate:* Albino rats were dark-adapted and sacrificed. The eyes were enucleated, the vitreous removed and the retina dissected under a dim red light. The retinas were washed and placed in the perfusion medium. The retinal tissue was maintained at 2 °C in all subsequent procedures except for the final incubation. The retinas were homogenized in 20 ml of 50 m*M* Tris-HCl buffer, pH 7.4, in a glass homogenizer with a teflon pestle and centrifuged at 48,000 *g* for 20 min. Discarding the supernatant, the pellet was resuspended in the same volume of buffer with an Ultra-Turax homogenizer and the homogenate was recentrifuged, washed, and resuspended in 20 ml of 20 m*M* Hepes, 0.32 *M* sucrose buffer, pH 7.4.

(2) *PAF-binding assay:* Aliquots of 50 µl of fresh membrane suspension (approx. 75 µg of protein) were used in binding experiments. The membrane suspension was added to 1.5 ml capacity Eppendorf capped tubes containing 0.25% BSA, 0.15 *M* NaCl, 50 m*M* Hepes buffer, pH 7.4 (total reaction volume was 500 µl). Following a short period of temperature equilibration (3 min at 37 °C), [^{3}H]PAF was added and the tubes incubated for 15 min at 37 °C. All incubations were terminated by centrifugation at 5,000 *g* for 5 min at 4 °C. Supernatants were removed and the retinal pellets were washed in buffer (1.2 ml) and recentrifuged in plastic tubes. Following this, the plastic tubes were cut 5 mm from the bottom and the section containing the pellet was placed directly in the scintillation bottle. The retinal pellets were digested in tissue solubilizer (Soluene-350; Packard, La Grange, Ill., USA) and neutralized with HCl before addition of scintillation fluid. Specific binding was calculated by subtracting nonspecific binding from total binding. Nonspecific binding was determined in the presence of 10^{-7} *M* of unlabelled PAF. All binding assays were performed in triplicate. Competition experiments were carried out on fresh retinal membrane preparations between [^{3}H]PAF (0.5 n*M*) and unlabelled PAF (10^{-10}; 10^{-7} *M*) or between [^{3}H]PAF (0.5 n*M*) and Ginkgolide B (BN 52021) (10^{-10}; 10^{-7} *M*) or WEB 2086 (3-[4-(2-chlorophenyl)-9-methyl-6H-thieno(3,2-b)(1,2,4)-triazolo(4,-3a) (1,4)-thienodiazepine-2-y1]-1-(4-morpholinyl)-1-propanone) (10^{-10}; 10^{-7} *M*).

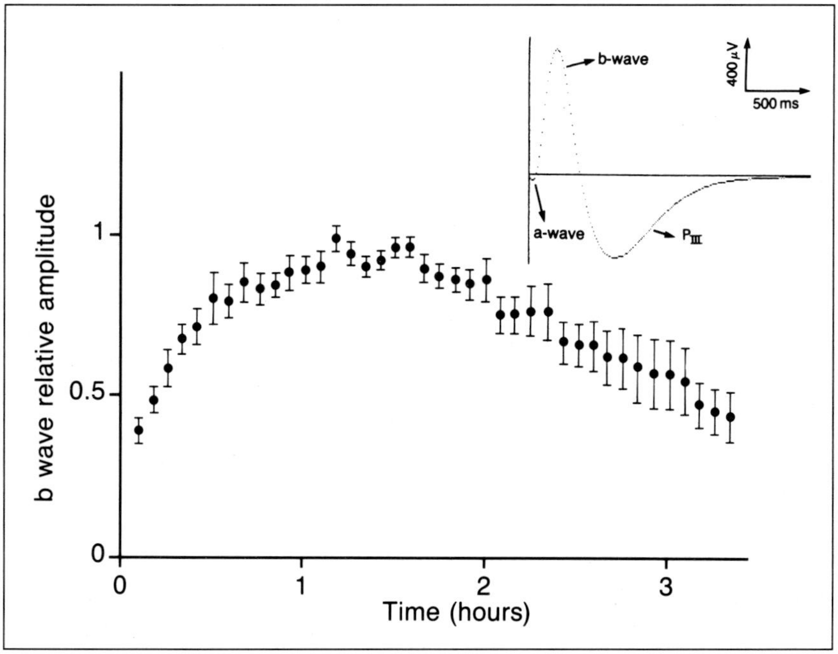

Fig. 2. Evolution of ERG b-wave amplitude recorded on rat isolated retina under normal conditions: white light stimulation, 300 lx, 1 ms (mean ± SEM; n = 10). Inset shows an example of ERG.

Results and Discussion

ERG of Isolated Rat Retina

Since the work of Ames and Gurian [14] and Weinstein et al. [15], we know that it is possible to maintain the function of an isolated retina, as monitored by its electrophysiological responses to light stimulation, for several hours. The variation of transretinal potential recorded when the isolated retina is white light stimulated (fig. 2) clearly shows that it contains rods almost exclusively. This signal is usually described as follows: at first a negative a-wave, then a positive and delayed b-wave, followed finally by a large, negative slow potential variation, the PIII process (according to Granit's nomenclature [16]). The intraretinal origin of these different waves is now well established [17]: the a-wave and PIII are generated almost only at the photoreceptor cells level, whereas the b-wave originates in a more internal cellular layer at the level of Müller's cells. This is why we chose to characterize the ERG quantitatively by measuring its b-wave amplitude, which best reflects the integrity of intraretinal nervous connections as well as the quality of tissue metabolism.

By systematically repeating the same stimulation at regular intervals (5 min in our experimental protocol) it appears that isolated retina provides average measurable responses ($> 10\ \mu V$) for about 4 h, and that curves of b-wave amplitude versus time all look the same (fig. 2): at the beginning, during the period when the retina gradually adapts to survival conditions, the b-wave amplitude first increases, then remains almost constant and finally decreases progressively until survival ends. This specific behavior led us to start any experiments only when the b-wave amplitude had reached its steady value, this value being taken as a reference [1].

Since the studies of Hagins [18], it is well known that the specific electrical signal given by photoreceptors, as a response to light stimulation, corresponds to the hyperpolarization of membrane of rod outer segments. This membrane hyperpolarization is related to a decrease of sodium permeability of the plasmic membrane (photocurrent). Thus stimulated, rods excite in their turn the other nervous cells of the retina (b-wave generation by Müller's cells) and the whole electrophysiological response of the retina constitutes the ERG. Furthermore, visible light detection at the retinal level is ensured by the rhodopsin molecule, which absorbs light photons. Rhodopsin is a protein which constitutes the saccules' membrane, these saccules being stacked up in a regular fashion in the rod outer segment. The first step of light detection is then purely photochemical (bleaching of the rhodopsin molecule), and the trigger of the electrophysiological signal (hyperpolarization of the rod outer segment membrane) involves the light activation of a phosphodiesterase which controls the cGMP plasmic concentration [19]. In its turn, cGMP regulates Na^+ permeability of rod outer segment membrane, variations in which generate the ERG [20].

Hydroxyl Radicals and Oxygenated Free Radical Scavengers

The results obtained with (Fe^{2+} + ascorbate) indicate that the survival of the retinas is considerably shortened when $OH^{\bullet}$ radicals are induced: it lasted approximately 1 h after induction of oxygenated free radicals whereas for the reference retinas the amplitude of the b-wave remains measurable for 4–5 h after the stationary phase (fig. 3). The mixture of Fe^{2+} + ascorbate provides an excellent peroxidative system of membrane lipids. Reacting with the membrane polyunsaturated fatty acids, the hydroxyl

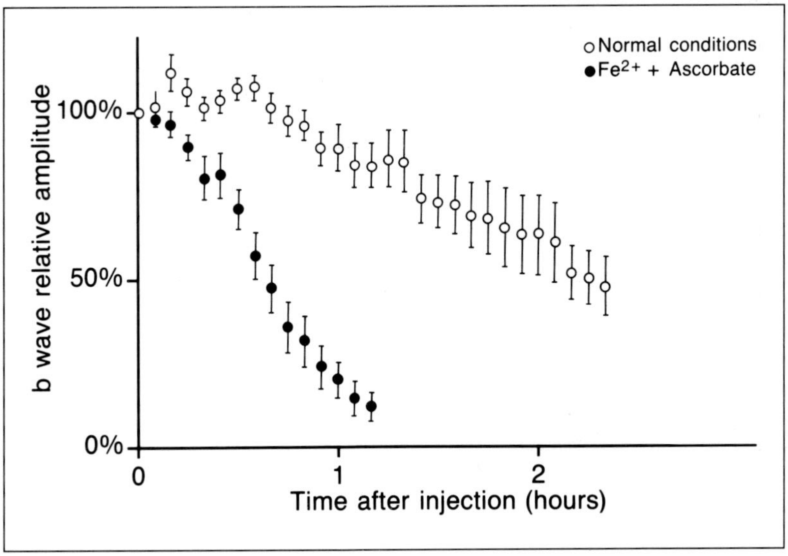

Fig. 3. Effects of OH˙ radicals induced by (Fe^{2+} + ascorbate) on the ERG amplitude (mean ± SEM).

radicals start off peroxidation chains with the intermediary formation of lipid radicals (L˙) or alcooperoxy ($L\dot{O}_2$). The latter two can therefore either (1) maintain the chain auto-oxidation reaction by reacting with a close fatty acids molecule or (2) decompose to fatty aldehydes at the origin of membrane polymerizations inhibiting enzymatic activities, or else (3) turn into hydroperoxide (LOOH) which is less reactive but capable of inducing a new series of membrane auto-oxidation processes.

Particularly rich in polyunsaturated fatty acids, the retina will be very sensitive to the reaction with lipoperoxidation products. Novikov et al. [21] have clearly established that the saccular membrane of receptors was particularly sensitive to lipid peroxidation: rhodopsin becomes more easily extractable due to a decrease in the lipid-protein interactions. In addition, according to Williams et al. [22], the kinetics of the conformational changes of rhodopsin during bleaching would also be affected by membrane lipoperoxidation. This particular sensitivity of the saccular membrane could therefore explain the decrease in the ERG amplitude.

When a free radical scavenger – as *G. biloba* extract – is added, the deleterious effects of lipoperoxidation are significantly reduced (fig. 4): the

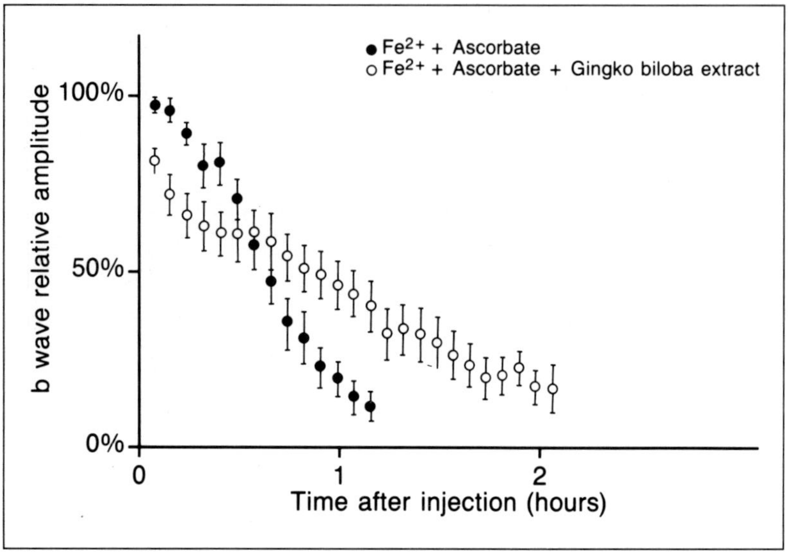

Fig. 4. Partial inhibition of retinal lipoperoxidation by addition of the *G. biloba* extract (mean ± SEM).

decreasing slope of b-wave amplitude is far less pronounced and the survival delay is practically increased twofold. The diminution of the ERG amplitude recorded just after addition of *G. biloba* extract can be attributed to the dark coloration of the perfusion solution [23]. Such a protective effect of *G. biloba* extract is directly connected to its free radical scavenger property. It so can be related to such molecules as α-tocopherol or butylhydroxytoluene and may play an important therapeutic role [24].

Leukotrienes

Addition of LTB_4, into the perfusion solution of the isolated retina, led to a significant decrease in ERG amplitude (fig. 5) compared to the average normal b-wave amplitude. The effect was dose-dependent: the drop in the b-wave was immediate with the highest doses used 10^{-7} *M* and 4×10^{-7} *M*) whereas the deleterious effect was slightly delayed with 2×10^{-8} *M*. After 1 h the b-wave amplitude reached about 50% of the control value for $LTB_4 = 2 \times 10^{-8}$ *M*, whereas the fall was more significant (20% of control) for $LTB_4 = 4 \times 10^{-7}$ *M*. After 2 h, the ERG signal had totally disappeared for the highest dose of LTB_4.

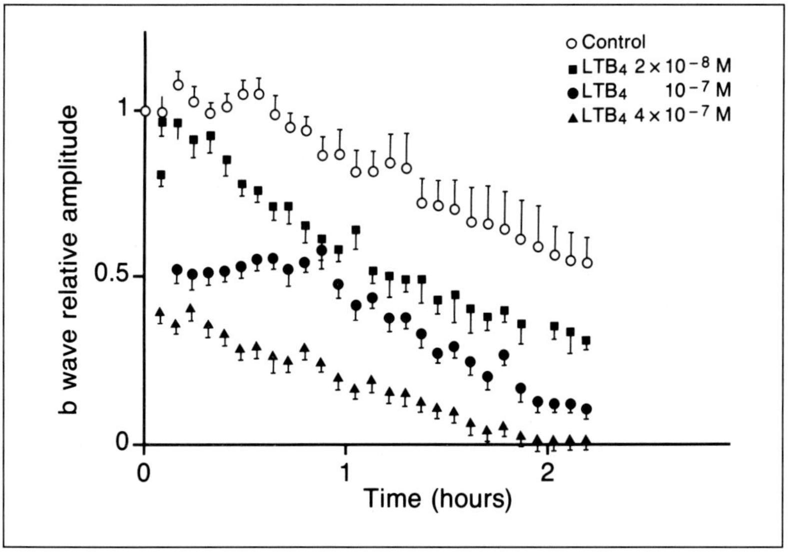

5

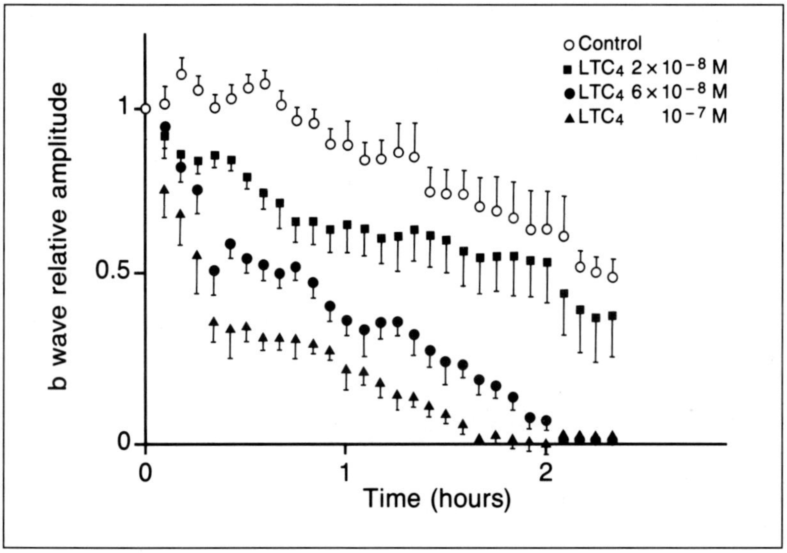

6

Fig. 5. Effects of leukotriene B_4 on the b-wave amplitude of ERG recorded on rat isolated retina.

Fig. 6. Effects of leukotriene C_4 on the b-wave amplitude of ERG recorded on rat isolated retina.

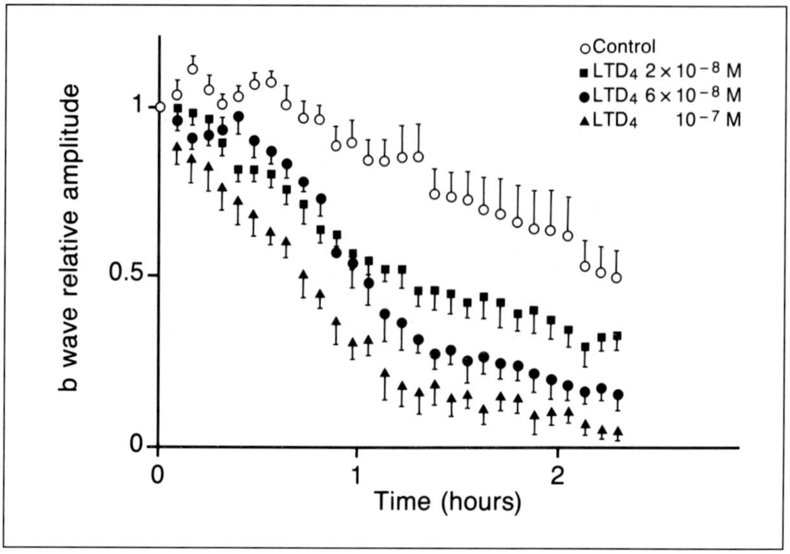

Fig. 7. Effects of leukotriene D_4 on the b-wave amplitude of ERG recorded on rat isolated retina.

The effect of LTC_4 was quite similar to that of LTB_4 (fig. 6). The drop in b-wave amplitude was dose-dependent, the maximal effect being observed with the highest dose used (10^{-7} *M*). In all cases, the initial decreasing slope was significantly pronounced in comparison with the normal curve. After 1 h the b-wave amplitude had significantly subsided about 60% of control for $LTC_4 = 2 \times 10^{-8}$ *M* and about 20% for $LTC_4 = 10^{-7}$ *M*. After 2 h the ERG signal had totally disappeared for the two highest doses of LTC_4.

Finally, the effect of LTD_4 was quantitatively comparable to that of LTB_4 (fig. 7). With the lower dose (2×10^{-8} *M*), the drop in ERG amplitude reaches near 50% after 1 h comparatively to the reference curve. This effect is dose-dependent involving significant differences in the decreases of b-wave amplitude.

These results indicate that with LTB_4 and LTD_4 the initial decreasing slope of the survival curve seems less pronounced than with LTC_4. In various experimental models, leukotrienes have been demonstrated to induce membrane permeability and gradient dissipation. Studies by Naccache et al. [5] indicate that LTB_4 provokes a rapid uptake of ^{45}Ca by neutrophils. This result was confirmed by Weissmann and Serhan [25] who demonstrated that only cis-LTB_4 and not its biologically inactive all-trans isomer, acted as a calcium ionophore in liposomes. A similar phenomenon has been demonstrated in both human erythrocytes [6] and mouse macrophages [26]. These changes in ionic membrane permeability triggered alterations of membrane potential: at the retinal level, we did not observe marked difference between LTB_4 and the peptidoleukotrienes C_4 and D_4, whereas at the neuronal level, Palmer et al. [27] observed only electrophysiological changes with LTC_4 and LTD_4. The differences in the experimental models are possibly the reason for these discrepancies. Another possible explanation for the inhibitory effect of leukotrienes on the retina is that the decrease in ERG amplitude could be related to the opposing effects of leukotrienes (depolarization) and light (hyperpolarization) at the level of the photoreceptor cell. This hypothesis has the advantage of taking into account the experimental results presented here, together with those reported in various studies with other models. Finally, our results showed that the decrease of amplitude of the b-wave precedes any modification of PIII. This reduction of the b-wave was progressive, dose-dependent and appeared to be characteristic of toxic effect [28] of leukotrienes on the retina.

Platelet-Activating Factor

When (R)PAF-acether (2×10^{-7} *M*) is added to the perfusion solution, ERG amplitude is markedly diminished compared to the reference curve (fig. 8). This fall is rapid and regular, and about 40 min after addition of (R)PAF-acether the amplitude decrease is more than 50%. In contrast, from 1 h after (R)PAF-acether addition until the end of survival, the ERG amplitude decreases slowly and the slope of the curve is nearly normal. This deleterious effect of (R)PAF-acether on the ERG is dose-dependent. Three different concentrations of (R)PAF-acether were tested: 2×10^{-11} *M*, 2×10^{-9} *M* and 2×10^{-7} *M*. As reported in figure 9, survival curves obtained indicate that the ERG amplitude decreases as a function of (R)PAF-acether concentration. Thus, about 40 min after (R)PAF-acether addition, b-wave amplitude reaches 74% of its reference value with 2×10^{-11} *M*, 48% with 2×10^{-9} *M* and 43% with 2×10^{-7} *M*. In contrast, when (S)PAF-acether (2×10^{-7} *M*) is used, no impairment of the profile of the survival curve is observed. Likewise, with lyso-PAF (2×10^{-7} *M*) the evolution of the b-wave amplitude remains almost identical to that recorded in control conditions. Finally, when (R)PAF-acether (2 ×

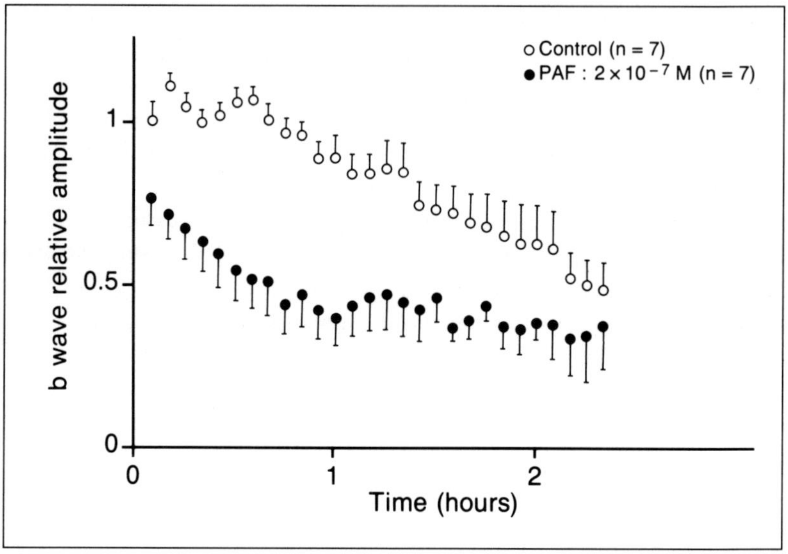

8

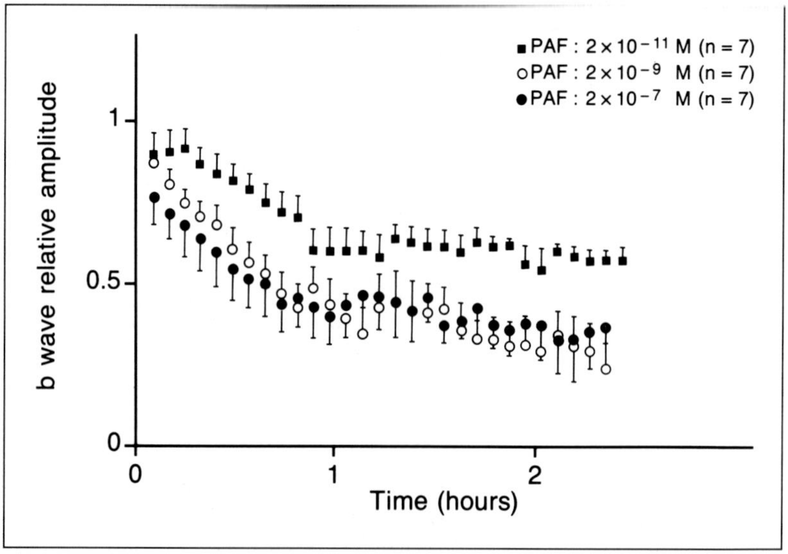

9

Fig. 8. Evolution of ERG b-wave amplitude in the presence of PAF (2×10^{-7} *M*) compared to the control curve.

Fig. 9. Dose-dependent effect of PAF on b-wave amplitude measured on isolated retina.

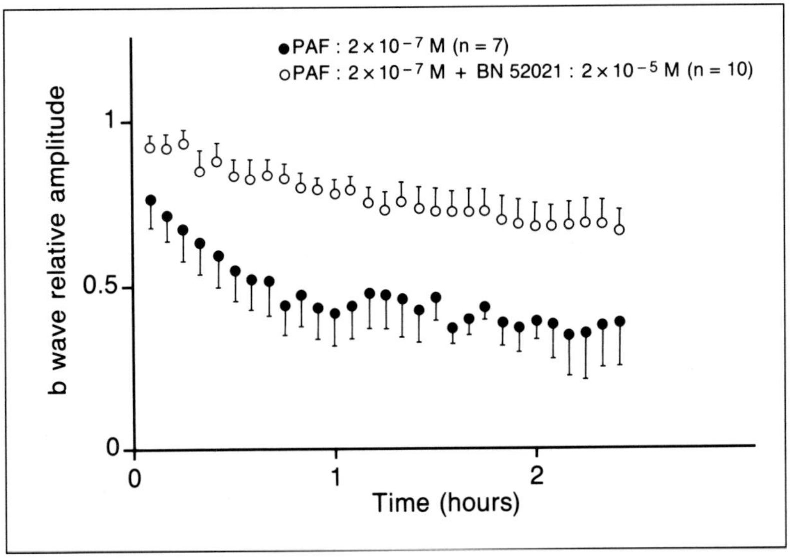

Fig. 10. Inhibition of PAF effects on ERG by simultaneous administration of BN 52021.

10^{-7} *M*) is added with Ginkgolide B (BN 52021: 2×10^{-5} *M*) the survival curve obtained is very comparable to the reference curve (fig. 10). After about 1 h of perfusion, the b-wave amplitude recorded with BN 52021 is nearly 90% of the reference value compared to 45% obtained with (R)PAF-acether alone. Furthermore, the evolution of b-wave amplitude in the presence of BN 52021 is perfectly normal until the end of the survival.

These results demonstrate that (R)PAF-acether, but not (S)PAF-acether or lyso-PAF, exerts a deleterious effect on the light response of vertebrate isolated retina and that these effects are dose-dependent. On other experimental models, it has been established that (R)PAF-acether binds to membrane-specific receptors and activates the phosphatidylinositol cycle with formation of inositol 1,4,5-triphosphate (IP_3), which in turn mobilizes intracellular Ca^{2+} [29, 30]. Therefore, (R)PAF-acether may be considered as a mediator using Ca^{2+} as messenger [reviewed in 31]. Such IP_3 formation might be the key of the (R)PAF-acether effects at the retinal level. Indeed, several recent papers [32–35] clearly establish that IP_3 is able to excite and light-adapt retinal photoreceptors, probably by raising

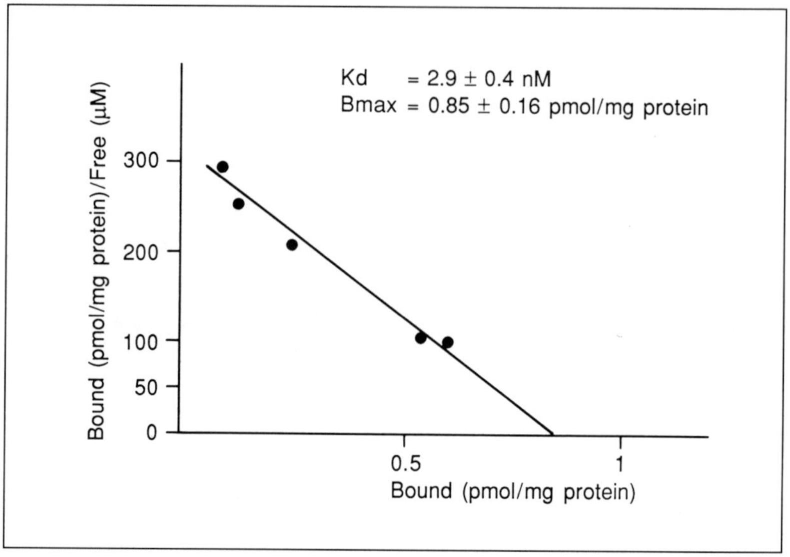

Fig. 11. Scatchard plot analysis of [^{3}H]PAF binding to rat retina homogenate.

[Ca^{2+}]$_i$. In these conditions, it can be assumed that ERG elicited on isolated retina after (R)PAF-acether application displays a decreasing amplitude corresponding to retinal light adaptation. Furthermore, it is interesting to note that (R)PAF-acether may be produced by chick retinas upon stimulation with various neurotransmitters; (R)PAF-acether produced is not released into supernatants and is extractable from retinas. In contrast, only a small amount of (R)PAF-acether is detectable following sonication of unstimulated retinas [36].

Of course, the hypothesis concerning the mechanism of (R)PAF-acether effects in the retina needed further confirmation, particularly concerning the existence of PAF receptors inside the retina.

Experiments performed on retinal membrane indicate that at 37 °C the binding of [^{3}H]PAF (0.5 n*M*) to the retinal homogenate was very rapid, binding being maximal by approximately 3 min and remaining stable for at least 30 min. The value of the half-equilibrium time was established as 45 s. The [^{3}H]PAF (0.5 n*M*) binding was a linear function of protein concentration over the range of 40–200 μg/ml and the specific binding represented 20–30% of the total binding. As shown in figure 11, at equilibrium

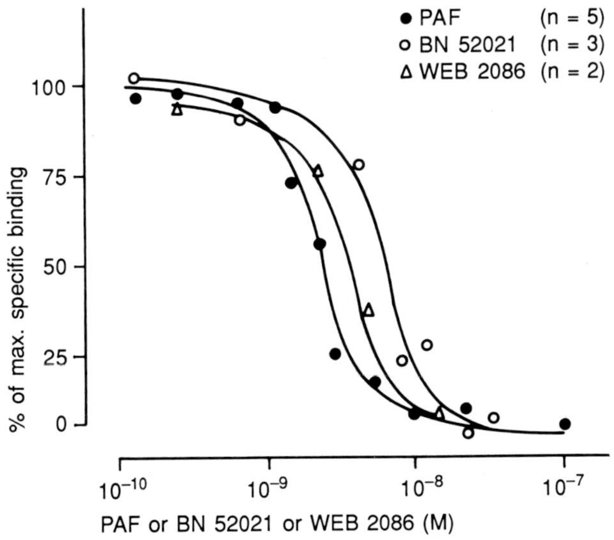

Fig. 12. Inhibition of specific [^{3}H]PAF binding to retina membrane preparations by increasing concentrations of PAF (10^{-10}; 10^{-7} *M*), BN 52021 (10^{-10}; 10^{-7} *M*) or WEB 2086 (10^{-10}; 10^{-7} *M*).

conditions, the saturation curve obtained from [^{3}H]PAF binding by Scatchard analysis indicated a high affinity binding site: $K_d = 2.9 \pm 0.4$ n*M*; $B_{max} = 0.85 \pm 0.16$ pmol/mg of protein. The specificity of PAF binding to retinal membrane was assessed by displacing [^{3}H]PAF with unlabelled PAF and the specific PAF receptor antagonists BN 52021 and WEB 2086. Figure 12 shows the displacement of the radioligand by unlabelled PAF, BN 52021 and WEB 2086. The inhibition constant (IC_{50}) of the specific binding of [^{3}H]PAF to retina membrane preparations was 2.1 n*M* as determined with unlabelled PAF. IC_{50} values of 5 and 3 n*M* were obtained with BN 52021 and WEB 2086 respectively.

The high affinity retinal site demonstrated by our experiments [37] has a K_d and a B_{max} comparable to the values proposed for the rabbit and human membranous models. In the latter, values in the range of 0.05–40 n*M* have been reported by various authors for the K_d of tritiated PAF [reviewed in 8]. In addition, the existence of two types of specific PAF-binding site were reported in the gerbil brain. These two classes of receptor site correspond to a $K_d = 3.66 \pm 0.92$ nM; $B_{max} = 0.83 \pm 0.23$ pmol/mg of membrane protein and a $K_d = 20.4 \pm 0.56$ nM; $B_{max} = 1.1 \pm 0.32$ pmol/mg of membrane protein, respectively for [^{3}H]PAF [38]. This finding is consistent with those reported in the present study, the K_d values being in the same order of magnitude as those of the high affinity retinal site. It should be noted, that due to its embryologic origins, the retina also constitutes an excellent brain model.

Conclusion

The ERG generation implicates very specific physicochemical and biochemical mechanisms; it corresponds to a succession of variations of membrane ionic permeability. Now, numerous inflammatory mediators develop a biological activity which expresses itself directly or not by modifying membrane ionic permeabilities. In these conditions, it seems essential to test these mediators on the retina because firstly the retina corresponds to an objective model allowing quantification of membrane impairments and secondly we are then able to specifically record the effects of these mediators on the visual process. Moreover, the retina being the neurosensorial tissue of the eye, it will be possible to predict the functional repercussions of ocular inflammatory diseases, in the course of which these mediators are released.

With our experimental model of isolated mammalian retina, we found again a particular sensitivity of the retina to the reactivity of hydroxyl radicals. Such free radicals, initiating membranous lipoperoxidation, are very reactive on the retina particularly rich in polyunsaturated fatty acids. So, we have been able to study the protective properties of radical scavengers especially of vegetal origin. Our results suggest that such radical scavengers may be used for ocular therapy when oxygenated free radicals are implicated – as for diabetic retinopathy for instance – and also perhaps when these radicals are involved in cerebral dysfunctions. Our experimentations performed with leukotrienes indicate that these molecules are able to induce a marked drop of ERG amplitude. When taking into account the known effects of these molecules such as the increase of Na^+ membrane permeability, we can suggest that leukotrienes and light develop opposite effects at the rod outer segment membrane level.

Finally, the results obtained with PAF demonstrate firstly the presence of specific PAF-binding sites inside the rat retina and secondly that these sites are implicated in visual function. Taking into account the consequences of the binding of PAF to its receptors on the phosphatidylinositol cycle, the fall in amplitude of the ERG induced by PAF could be due either to a partial loss of adaptation of the retina to dark or to a desensitization of the photoreceptor cell linked with the mobilization of intracellular Ca^{2+}. Thus it appears that PAF may play the role of a functional mediator in the retina, modulating the response to light and possibly being operative in inflammatory disorders of visual function.

References

1 Doly M, Braquet P, Bonhomme B, et al: Effects of lipid peroxidation on the isolated rat retina. Ophthalmic Res 1984;16:292.
2 Kagan VY, Shvedova AA, Novikov KN, et al: Spontaneous and induced autooxidation of phospholipids during conformational rearrangements of the membrane of the external segments of the frog retinal rods. Biofizika 1975;20:1043.
3 Anderson RE, Sperling L: Lipids of ocular tissues. VII. Positional distribution of the fatty acids in the phospholipids of bovine retina rod outer segments. Arch Biochem Biophys 1971;144:673.
4 Borgeat P, Samuelsson B: Metabolism of arachidonic acid in polymorphonuclear leucocytes: unstable intermediate in formation of dihydroxy acids. Proc Natl Acad Sci USA 1979;76:3213.
5 Naccache PH, Sha'afi RI, Borgeat P, et al: Mono- and dihydroxyeicosatetraenoic acids after calcium homeostasis in rabbit neutrophils. J Clin Invest 1981;67:1584.
6 Garay RP, Diez J, Braquet P: Are icosanoids involved in ion transport regulation? I. A study in human red blood cells; in Braquet P, Garay RP, Frolich JC et al (eds): Prostaglandins and Membrane Ion Transport. New York, Raven Press, 1984, vol 1, pp 135–146.
7 Benveniste J, Leconédic JP, Polonski J, et al: Structural analysis of purified platelet-activating factor by lipases. Nature 1977;269:170.
8 Braquet P, Touqui L, Shen TY, et al: Perspectives in platelet-activating factor research. Pharmacol Rev 1987;39:97.
9 Garay RP, Braquet P: Involvement of K^+ movements in the membrane signal induced by PAF-acether. Biochem Pharmacol 1986;35:2811.
10 Rodieck RW: The Vertebrate Retina: Principles of Structure and Function. San Francisco, Freeman, 1973.
11 Arden GB, Ernst W: The effect of ions on the photoresponses of pigeon cones. J Physiol (Lond) 1970;211:311.
12 Doly M, Isabelle DB, Vincent P, et al: Mechanism of the formation of X-ray induced phosphenes. I. Electrophysiological investigations. Rad Res 1980;82:93.
13 Braquet P: Treatment or prevention of PAF-acether disorders provoked by a new

series of highly specific inhibitors; GB Patent, 84/18, 424 (19 July 1984) Belg. BE 901, 915 (see CA 103: 189 808d, 1985).
14 Ames A, Gurian BS: Electrical recordings from isolated mammalian retina mounted as a membrane. Arch Ophthalmol 1963;70:139.
15 Weinstein GW, Hobson RR, Dowling JE: Light and dark adaptation in the isolated rat retina. Nature 1967;215:134.
16 Granit R: The components of the retinal action potential in mammals and their relation to the discharge in the optic nerve. J Physiol 1933;77:297.
17 Tomita T, Yanagida T: Origins to the ERG waves. Vision Res 1981;21:1703.
18 Hagins WA: The visual process: excitatory mechanisms in the primary receptor cells. Annu Rev Biophys Bioeng 1972;1:131.
19 Chabre M: Trigger and amplification mechanisms in visual phototransduction. Annu Rev Biophys Biophys Chem 1985;14:331.
20 Stein PJ, Rasenick MM, Bitensky MW: Biochemistry of the cyclic nucleotide-related enzymes in rod photoreceptors. Retinal Res 1982;1:227.
21 Novikov KN, Kagan VY, Shvedova AA, et al: Protein-lipid interactions on peroxidative oxidation of lipids in the photoreceptor membrane. Biofizika 1975;20: 1058.
22 Williams TP, Baker BN, McDowell D: The influence of lipids on dynamic properties of rhodopsin. Exp Eye Res 1974;18:69.
23 Doly M, Braquet P, Drieu K, et al: Protecting effects of *Ginkgo biloba* extract on the retinal function; in Verriest G (ed): Colour Vision Deficiencies. VIII. Dordrecht, Nijhoff & Junk, 1987, pp 515–523.
24 Doly M, Braquet P, Droy MT, et al: Alteration of electrophysiological function of isolated retina from alloxan-induced diabetic rats: effect of treatment with *Ginkgo biloba* extract. Neurochem Pathol 1988;8:15.
25 Weissmann F, Serhan C: Leukotriene B_4 and phosphatidic acid are calcium ionophores; in Braquet P, Garay RP, Frolich JC, et al (eds): Prostaglandins and Membrane Ion Transport. New York, Raven Press, 1984, vol 1, pp 129–134.
26 Diez J, Braquet P, Nazaret C, et al: The effect of cyclic nucleotides and eicosanoids on Na^+ and K^+ transport systems in human red cells and mouse macrophages; in Greengard P, Robinson GA (eds): Advances in Cyclic Nucleotide and Protein Phosphorylation Research. New York, Raven Press, 1984, vol 17, pp 1–615.
27 Palmer MR, Mathews WR, Hoffer BJ, et al: Electrophysiological response of cerebellar Purkinje neurons to leukotriene D_4 and B_4. J Pharmacol Exp Ther 1981;219: 91.
28 Honda Y: The mode of action of neurotoxins and some common chemicals upon the electrophysiological properties of perfused mammalian retinas. Doc Ophthalmol 1978;15:3.
29 Lapetina EG: Platelet-activating factor stimulates the phosphatidylinositol cycle. J Biol Chem 1975;250:4427.
30 Billah MM, Lapetina EG: Platelet-activating factor stimulates metabolism of phosphoinositides in horse platelets: possible relationship to Ca^{2+} mobilization during stimulation. Proc Natl Acad Sci USA 1983;80:965.
31 Berridge MJ: Inositol triphosphate and diacylglycerol as second messengers. Biochem J 1984;220:345.

32 Brown JE, Rubin LJ, Chalagini AF, et al: Myoinositol polyphosphate may be a messenger for visual excitation in *Limulus* photoreceptors. Nature 1984;311:160.

33 Fein A, Pagne R, Corson DW, et al: Photoreceptor excitation and adaptation by inositol 1,4,5-triphosphate. Nature 1984;311:157.

34 Birkle DL, Bazan NG: The arachidonic acid cascade and phospholipid and docosahexaenoic acid metabolism in the retina. Retinal Res 1986;5:309.

35 Pagne R, Corson DW, Fein A, et al: Excitation and adaptation of *Limulus* ventral photoreceptors by inositol 1,4,5-triphosphate result from a rise in intracellular calcium. J Gen Physiol 1986;88:127.

36 Bussolino F, Gremo F, Tetta C, et al: Production of platelet-activating factor by chick retina. J Biol Chem 1986;261:16502.

37 Thierry A, Doly M, Braquet P, et al: Presence of specific platelet-activating factor binding sites in the rat retina. Eur J Pharmacol, 1989;163:97.

38 Domingo MT, Spinnewyn B, Chabrier PE, et al: Presence of specific binding sites for platelet-activating factor in brain. Biochem Biophys Res Commun 1988;151:730.

Michel Doly, PhD, Laboratoire de Biophysique, Facultés de Médicine et de Pharmacie, Unité INSERM U.71, F–63000 Clermont-Ferrand (France)

Bazan NG (ed): Lipid Mediators in Eye Inflammation.
New Trends Lipid Mediators Res. Basel, Karger, 1990, vol 5, pp 168–180

Ocular Responses to Inflammation and the Triggering of Wound Healing: Lipid Mediators, Proto-Oncogenes, Gene Expression, and Neuromodulation

Nicolas G. Bazan, Haydee E.P. Bazan

LSU Eye Center and Neuroscience Center, New Orleans, La., USA

As has been described through the various chapters of this book, the inflammatory response differs throughout the eye, however, in every tissue, the basic events comprise an initial response by resident cells associated with the arrival of inflammatory cells, mainly polymorphonuclear leukocytes (PMNs), followed by the appearance of mononuclear cells, and then proliferation of fibroblasts. The accumulation of these cells at sites of inflammation is required for elimination of microbial agents and/or cellular debris, in order to prepare the tissue for the repair process (fig. 1).

In experimental uveitis, for example, there is a rapid appearance of PMNs in the iris and ciliary bodies, followed by a prolonged presence of mononuclear cells (1–14 days) in the stroma of the ciliary bodies as well as in the iris [1]. The responses to proinflammatory stimuli are complex and also involve the immune system, mainly T lymphocytes. In some instances, such as in various forms of uveitis, sustained T cell activation, impaired phagocytosis, and events not well understood take place, and chronic forms of inflammation ensue (fig. 1).

Others factors can also delay the process of normal wound healing, such as accumulation of free radicals by phagocytic cells. PMNs generate superoxide radicals ($O_2^{\cdot}$) and hydrogen peroxide (H_2O_2) as their oxidative metabolism increases ('respiratory burst') to initiate their bactericidal action. Cells normally contain antioxidant mechanisms (e.g. superoxide dismutase, glutathione, ascorbic acid, vitamin E) that counteract free radical-induced cell damage. However, if the production of free radicals

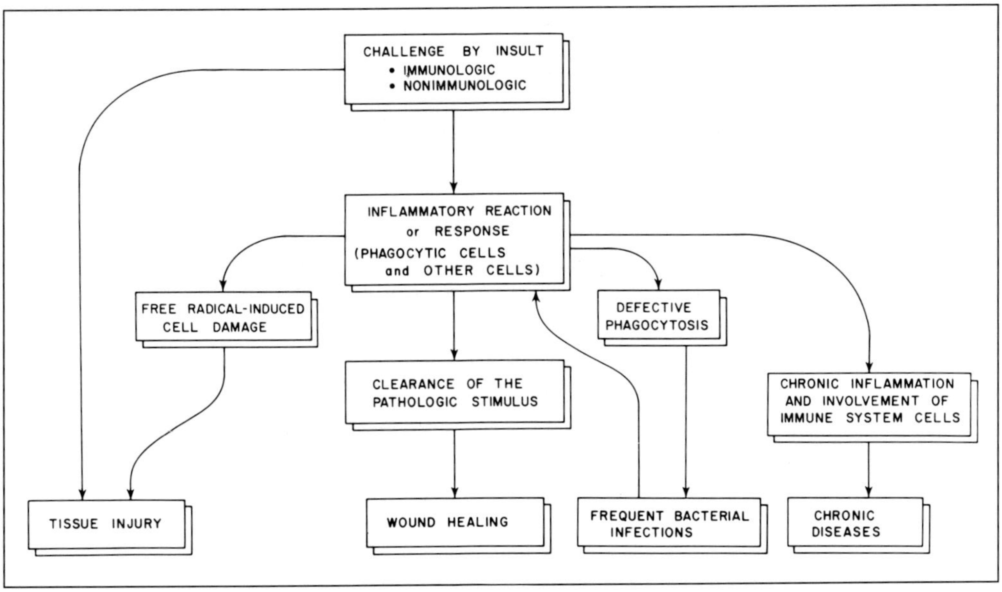

Fig. 1. Outline of the inflammatory response. Diverse challenges promote cell activation and recruitment of phagocytes. Thereafter, free-radical induced cellular damage and tissue injury may arise. Alternatively the triggering insult may be removed or isolated and cell repair and wound healing may occur. Chronic inflammatory conditions arise particularly when there is an involvement of the immune system (e.g. uveitis) with or without impairment of phagocytosis.

exceeds the capacity of the antioxidant defense mechanisms, tissue damage takes place. The mechanisms of cell damage by free radicals are complex. Free radicals can inactivate enzymes, stimulate the process of lipid peroxidation and lead to DNA fragmentation and mutations.

The inflammatory response comprises, in addition, several chemical mediators, including oxygenated metabolites of arachidonic acid (20:4) (e.g. prostaglandins, thromboxanes, hydroxyeicosatetraenoic acids, leukotrienes), cytokines (e.g. interleukins, cytotoxins), platelet-activating factor, and complement components, which in turn elicit effects on local vasculature, immunoprocesses and cell function.

The consequences of the inflammatory response are therefore the result of an interplay between networks of chemical mediators and cell-cell interactions. In various parts of the eye, such as the iris and cornea, interactions between nerve cell terminals and nonneural cells also play a fundamental role in the outcome of inflammation.

Many of the chemical mediators accumulated are second messengers, generated by receptor activation under physiological conditions. A better knowledge of these cell-signal transduction events may contribute to an understanding of the molecular bases of the coupling of inflammatory responses to triggering of events leading to cell survival, repair, proliferation and wound healing. Cell-signaling events may lead to transcriptional modulation, either positive or negative, of genes that in turn are engaged in cell repair and wound healing. In this chapter we summarize aspects of our work relevant to lipid mediators in eye inflammation. We also discuss our hypothetical model, mainly related to corneal inflammatory responses and the triggering of wound healing, based on lipid mediators, gene expression, and the modulatory nature of the sensory innervation.

Receptor, G Proteins, Phospholipases, and the Arachidonic Acid Cascade

Among the different mediators involved in the eye inflammatory process, the arachidonic acid cascade metabolites are the most intensely studied. These metabolites are oxygenated derivatives synthesized through enzyme-mediated steps that result in biologically active reaction products. For example, certain prostaglandins reduce intraocular pressure and may modulate ocular inflammation [2], while leukotriene B_4 (LTB_4) recruits PMNs in the anterior chamber and tissues [3, 4]. In eye tissues, such as anterior uvea [5, 6], conjunctiva [5, 6], retina [7], iris [8], retinal pigment epithelium [9] and cornea [10–12], the ability to synthesize 20:4 metabolites has been studied.

Inflammatory responses occur rapidly due to activation of resident cells as a first step in the response. Figure 2 is a composite of cellular events

---

Fig. 2. Diagrammatic representation of the possible events coupling inflammation with cell repair and wound healing. The events depicted here are a composite of those taking place in such diverse cells as epithelial, endothelial and neural. The cell-signal transduction molecules depicted, on the left of the diagram, on the plasma membrane lead to the formation of second messengers (mediators) that in turn bring about short-term re-

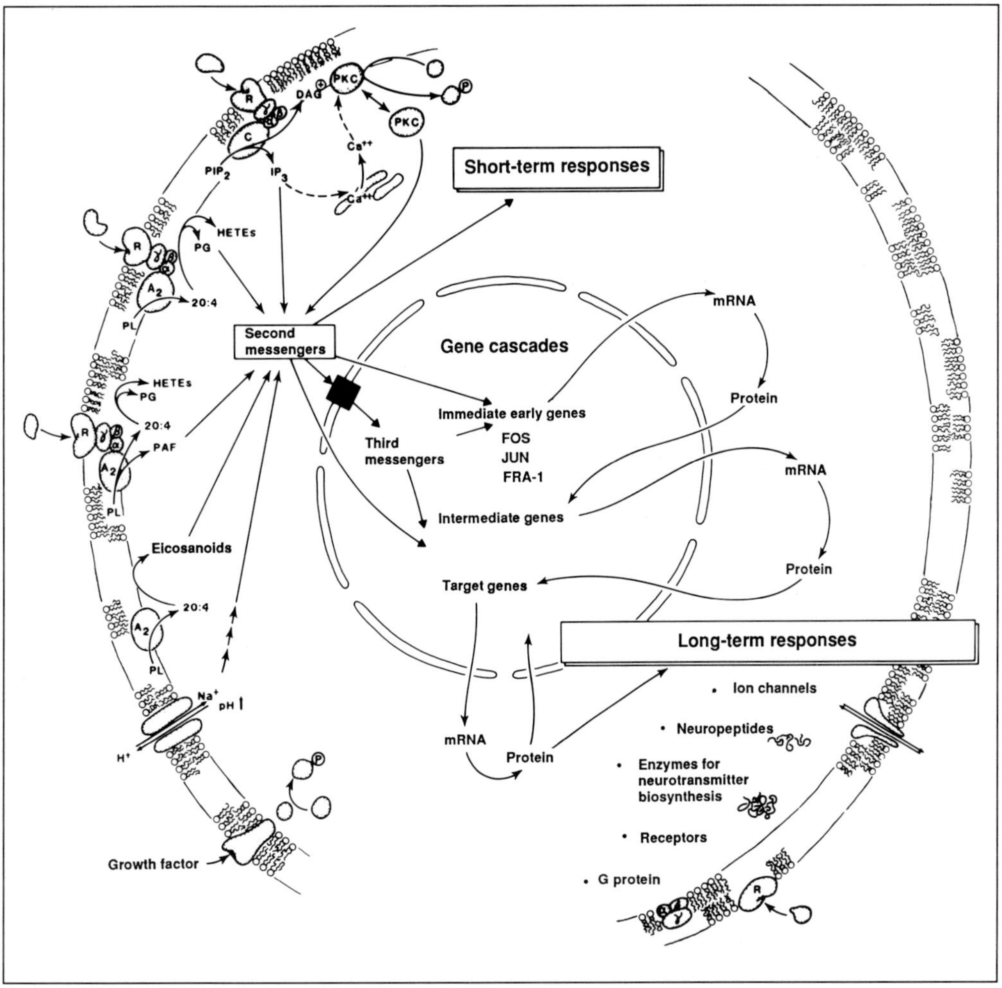

sponses. These second messengers or mediators are synthesized upon the interaction of an agonist (e.g. hormone, growth factor, neurotransmitter) with a receptor and in turn elicit a physiological response. In inflammation several of these messengers or mediators are accumulated and actively participate in the pathophysiological responses. In the center of the figure, gene cascades are outlined starting from immediate-early genes and possible effects of the second messengers are indicated. Target genes then generate long-term responses by synthesizing mRNA that encodes proteins involved in cell repair. How the events are interlinked in a sequential manner is not understood. Several of the events excluding synthesis of neuropeptides and enzymes for neurotransmitter biosynthesis occur in ocular non-neural cells. However, information in these cells is rather scarce.

involved in inflammation as well as in the initiation of the repair process. Several events, involving cell-signal transduction across the plasma membrane and leading to the formation of second and third messengers inside the cell, are indicated. This model implies that these intracellular messengers will in turn elicit modulatory effects on gene expression. These nuclear events then trigger the initial cellular changes after injury, which initiate wound healing.

Prostaglandins, leukotrienes and other eicosanoids are formed and released upon stimulation. The common precursor 20:4 is stored esterified on C_2 of the glycerol backbone of cellular phospholipids and is released by the activation of either of two enzymes: phospholipase A_2 (A_2 in fig. 2), which hydrolyzes 20:4 from phospholipids such as phosphatidylcholine; or phospholiapse C (C in fig. 2), which acts on inositol lipids. When the substrate for phospholipase C is phosphatidylinositol-4,5-bisphosphate (PIP_2), the products are diacylglycerol (DAG) and inositol triphosphate (IP_3). DAG is also the product when the substrate of phospholipase C is phosphatidylinositol or phosphatidylinositol-4-phosphate. DAG is an intracellular messenger that in turn activates protein kinase C. DAG can be subsequently deacylated by a lipase, releasing 20:4. The other product of phospholipase C, IP_3, is an intracellular mediator that triggers Ca^{2+} mobilization from nonmitochondrial stores. It seems that more than one storage site is involved [13].

Some of the phospholipases involved in the release of 20:4 are regulated by guanine nucleotide regulatory proteins (G proteins) [14, 15]. G proteins are a modulatory link between agonist and effector of the stimuli. They interact with the membrane receptors for extracellular signals such as leukotrienes, hormones, and photons of light (in the case of transducin in retinal photoreceptors). These G proteins are composed of three subunits: α, β, γ. The α subunits give specificity to the different G proteins. When activated, the G proteins bind to GTP and dissociate the α chain from the β, γ subunits. The released Gα-GTP then activates a target enzyme (A_2 and C in fig. 2) which releases 20:4 [14].

Free 20:4 thus generated can be used as substrate for the membrane-bound cyclooxygenase, giving rise to prostaglandins (PG) PGE_2, PGD_2, $PGF_{2\alpha}$, prostacyclin and thromboxane. Alternatively, one of several lipoxygenases may convert free 20:4 into hydroxy fatty acids (HETEs), some of which in subsequent steps are substrates for leukotrienes B_4, C_4, D_4, and E_4. Also di- or trihydroxylated derivatives, such as lipoxins or epoxilins, are components of the 20:4 cascade.

Platelet-Activating Factor

Another pathway activated during injury is the synthesis of platelet-activating factor (PAF). This phospholipid is a potent mediator of inflammatory and immune reactions [16], and increases after injury in the cornea [17]. The activation of the PAF cycle requires the stimulation of a phospholipase A_2 (fig. 2), which acts on an inactive membrane component precursor of PAF (alkyl-arachidonoyl-glycerophosphocholine). Topical application of a PAF antagonist inhibits the appearance of lipoxygenase reaction products in the cornea and aqueous humor after alkali-burn injury to the cornea [18–20]. This demonstrates that, in the cornea, the formation of PAF after injury contributes to the synthesis of other inflammatory mediators, such as eicosanoids. In an experimental model of keratitis and uveitis, topical application of two PAF antagonists, BN 50726 and BN 50730, delays the appearance of corneal epithelial damage, exudation and iris erythema up to 4 days after onset [unpublished data from our laboratory].

Modulation of Cellular Growth, Growth Factors, and Angiogenesis

The elucidation of the molecular and pharmacological bases of the control of cellular proliferation seems a necessary first step in the development of therapeutically useful agents for the management of corneal epithelial defects and incomplete corneal wound closures. Much of the emphasis to date has been on peptide growth factors (e.g., EGF, bFGF, aFGF) and newly recognized ones such as platelet-derived growth factor [21].

The lipid mediator DAG is generated in corneal endothelial cells exposed in culture to EGF. Since phospholipase C seems to be activated, as judged by the increases in IP_3 [22], it seems, therefore, that at least some of the growth factors have receptors coupled to enzymes that give rise to lipid mediators in the eye. Among the problems in the study and understanding of the interaction of growth factors in the eye is that, in addition to promoting wound healing, they also stimulate angiogenesis, an undesirable side effect. An alternative approach is the study of endogenous modulators of cellular proliferation and of wound healing. This approach may take two directions. First is the study of the effects of lipid mediators on cellular proliferation. We know that PAF, one of the mediators accumulated in the cornea after injury [17], does activate transcription [23–25]. Therefore, it is conceivable that some of the lipid mediators generated during inflammation may contribute to the turning-on of cellular proliferation. Secondly, muscarinic cholinergic receptors ($mACh_1$ and $mACh_4$), coupled through phospholipase C hydrolysis of PIP_2, give rise to IP_3 and to the lipid mediator DAG, as already mentioned. Agonists acting through this mechanism have been shown to stimulate DNA replication and hence cellular proliferation [26]. One of the salient features of the cornea is its very high content of acetylcholine as well as all cholinergic metabolizing enzymes [27, 28]. However, appropriate receptors have not been detected. Perhaps subtypes of the muscarinic cholinergic receptors should be explored. A possible role for the high concentration of acetylcholine in the cornea may be that of an endogenous mitogenic neurotransmitter contributing to the triggering of wound healing through the lipid mediators generated from inositol lipids.

Na^+/H^+ Antiport and Localized Cytoplasmic Alkalinization

Other mechanisms by which second messengers can be generated is by a localized alkalization of the cytoplasm. This change in pH is brought about by an agonist that activates the Na^+/H^+ antiport [29]. In platelets, thromboxane, ADP and PAF all accelerate Na^+/H^+ exchange [30] and may in turn activate a phospholipase A_2 [31], producing a cascade of lipid mediators.

Free Radical Cell Damage in the Eye

Inflammation in the eye also promotes the generation of free radicals. Recruited inflammatory cells (e.g. PMN and macrophages), necessary for ingestion and killing of microorganisms and cellular debris, generate free radicals. These cells arrive at the site of injury in response to chemoattractants such as complement protein C_{5A}, LTB_4, interleukin-1 and others. For example, in alkali-burned corneas, a leukocyte chemotactic factor that is not LTB_2 has been found [32]. Chemoattractants are also involved in the stimulation of phagocytosis, producing an oxidative burst that generates several forms of reduced oxygen (e.g. superoxide, hydrogen peroxide), which are bactericidal and promote further tissue inflammation and cellular damage through the activation of proteases [33, 34]. Cellular proliferation (such as occurs after wounding of the cornea or injury to other parts of the eye) is susceptible to DNA damage and mutation by oxygen radicals [35].

The Coupling of the Inflammatory Response to Wound Healing: The Activation of Gene Cascades

Are second messengers engaged in the communication with the nuclei to promote cellular repair and wound healing? Lipid mediators may be involved in the coupling of short-term responses to genomic events, leading to long-term cellular modifications. Figure 2 illustrates the following possible roles of some second messengers: (a) direct transcriptional activation; (b) eliciting of the synthesis of third messengers (black box of figure 2) that in turn promote gene expression. Receptor-mediated events at the plasma membrane may also subsequently activate DNA replication and cellular proliferation (not shown). These events will be discussed for corneal wound healing, although relevant information is just emerging. The expression of *c-myc* [36] and *c-fos* [unpublished data from our laboratory] has been detected in the rabbit cornea. Moreover, *c-fos* transcriptional activation seems to be enhanced soon after experimental alkali burn [unpublished results from our laboratory]. *c-fos* is a proto-oncogene, also called immediate-early gene, because it is rapidly transcribed upon signaling by a number of cell activators. As depicted in figure 2, second messengers may directly or indirectly exert an effect on these genes. In the injured cornea there may be modulators of gene expression among the lipid mediators accumulated, including PAF. As already mentioned, PAF acts as a transcriptional activator of *c-fos* and *c-jun* in neuroblastoma cells [25]. These early activated genes may in turn lead to transcriptional enhancement of gene cascades, through intermediate genes and then through target genes that encode proteins necessary for the long-term response. In neuroblastoma cells it was also shown that gene cascades may be turned on by the lipid mediator PAF, since the *c-fos* transcriptional activation is followed by an increase in mRNA for calcyclin, a Ca^{2+}-binding protein, but not of mRNA for HSP (heat-shock proteins) [37].

In conclusion, figure 2 illustrates our hypothesis on the coupling of the inflammatory response to the long-term genomic events that lead to the transcriptional modulation underlying pseudopod formation, cellular migration, sliding, and other mechanisms for covering the wound, and thereafter DNA replication and cellular proliferation.

Neuromodulation of Corneal Function and Wound Healing

There is ample evidence for the functional relevance of the interactions between the nervous system and ocular tissues, e.g. the control of the synthesis of lipid mediators in the iris is brought about by neurotransmitters released by the nerves [8].

The corneal epithelium and stroma contain free nerve endings and represent the most enriched sensory innervation of all surface epithelium [38]. These nerve fibers are from the long ciliary nerves, branches of the ophthalmic nerve derived from the trigeminal unipolar neurons. The terminals of these neurons display a uniquely rapid remodeling in vivo [39]. Neurotrophic actions on the maintenance of corneal properties as well as in wound healing and during aging are well known [40–42]. Much can be gained by a molecular neurobiological approach on the regulation of gene expression in the trigeminal bipolar neurons that innervate the cornea under normal conditions and during corneal injury.

The transcription of a gene that encodes for a nerve terminal protein (GAP-43, which is also called F_1, B-50, and ssp63) is enhanced during the development of growth cones in the nervous system, as well as in the mature tissue when sprouting and/or nerve regeneration takes place, such as during long-term synaptic potentiation [43], after ischemic damage [44], and in several other situations when plasticity and remodeling of nerve endings takes place. Although the expression of this gene and the presence of GAP-43 protein are currently being studied in the cornea and trigeminal ganglia [unpublished studies from our laboratory], protein kinase C is present in the corneal epithelial layers [45]. It has already been demonstrated in other tissues that GAP-43 is phosphorylated by protein kinase C [46].

Other components of growth cones and of mature nerve plasma membranes are phospholipids with large amounts of docosahexaenoyl acyl chains [R. Martin and N.G. Bazan, unpubl. data]. This fatty acid belongs to the essential fatty acid family of linolenic acid (omega-3 or n-3) and is also present in phospholipids of rabbit cornea [47]. There is no direct evidence that either the protein kinase C or 22:6-containing phospholipids present in cornea are in fact components of nerve endings of the cornea. Protein kinase C, although expressed in the highest amounts in the nervous system, particularly in nerve terminals [48], is also present in many other cells. By far the greatest enrichment of 22:6 is, however, in cellular membranes of the nervous system [49]. The synthesis of 22:6-containing phospholipids takes place in the endoplasmic reticulum, and it requires the turning-on of specific metabolic steps to supply this lipid to the plasma membrane of corneal sensory nerves after injury. Understanding of the events that signal the need for replenishing of both structural and functional molecules of the nerve endings of the cornea to the cell body of the bipolar trigeminal neurons will contribute to the development of novel therapeutic strategies to promote corneal healing. For example, recurrence of herpetic keratitis involves the harboring of herpes simplex virus in trigeminal nerves, and neural changes are related to the severity of the infection [50]. Nerve degeneration in this case may be associated with an interference in the transcriptional activation of mRNA processing of genes that are expressed in the trigeminal ganglia and needed for the maintenance of the functional and structural integrity of the cornea sensory innervation. Many new and exciting experiments are waiting to give us a better understanding of the complex events that initiate inflammation and are followed by the process of wound healing.

Conclusion

We have put forward a hypothesis linking lipid mediators and other events of inflammation with long-term responses at the genomic level. The outcome of this coupling is the modulation of cell repair and wound healing.

The progress made in recent years to understand the mechanisms of eye inflammation has contributed to the definition of cellular involvement and intracellular messengers (mediators), receptors, signal transduction pathways and mechanisms of regulation. This chapter is centered on a hypothetical scheme of how external events (stimuli) are connected to membrane events, intracellular events, and finally, nuclear events that result in the wound healing process. The description is based mainly on corneal tissue and on data from our laboratories. Some of the ideas described are waiting to be experimentally corroborated. With a molecular biological approach, the advances in this area, no doubt, will produce a series of exciting results that will bring more detail to the complex process of inflammation and tissue repair.

The progress made in recent years toward an understanding of the ocular response to inflammation has led to the elucidation of the cellular and molecular events comprising a network of chemical mediators. Lipid mediators, accumulated in the eye in response to injury, including immune challenges (e.g. uveitis), stand out as important possible targets for drug manipulation. The use of selective synthetic inhibitors and/or receptor antagonists for lipid mediators is opening up new avenues of pharmacological exploration and revealing possible therapeutic applications.

References

1 Uusitalo H: An experimental uveitis induced by bovine serum albumin. A transmission and scanning electron microscopic study. Acta Ophthalmol 1984;62:1–12.

2 Bito LZ: Prostaglandins and other eicosanoids: Their ocular transport, pharmacokinetics and therapeutic effects. Trans Ophthalmol Soc UK 1986;105:162–170.

3 Bhattacherjee P, Hammond B, Salmon JA, et al: Chemotactic response to some arachidonic acid lipoxygenase products in the rabbit eye. Eur J Pharmacol 1981;73: 21–28.

4 Sjternschantz J, Sherk T, Sears ML: Ocular responses to leukotriene C_4 and D_4 in the cat. Prostaglandins 1984;27:5–16.

5 Kulkarni PS, Fleisher L, Srinivasan BD: The synthesis of cyclooxygenase products in ocular tissues of various species. Curr Eye Res 1984;3:447–452.

6 Kulkarni PS, Kaufman PL, Srinivasan BD: Cyclooxygenase and lipoxygenase pathways in cynomolgus and rhesus monkey conjunctiva, anterior uvea and eyelids. Curr Eye Res 1987;6:801–808.

7 Birkle DL, Bazan NG: Lipoxygenase and cyclooxygenase reaction products and incorporation into glycerolipids of radiolabelled arachidonic acid in the bovine retina. Prostaglandins 1984;27:203–216.

8 Abel Latif AA: Regulation of arachidonate release, prostaglandins synthesis, and sphincter constriction in the mammalian iris-ciliary body; in Bito LZ, Stjernschantz J (eds): The Ocular Effects of Prostaglandins and Other Eicosanoids. New York, Liss, 1989, pp 53–72.

9 Bazan NG, Bazan HEP, Birkle DL, et al: Synthesis of leukotrienes in the frog retina and retinal pigment epithelium. J Neurosci Res 1987;18:591–596.

10 Bazan HEP, Birkle DL, Beuerman R, et al: Cryogenic lesion alters the metabolism of arachidonic acid in rabbit cornea layers. Invest Ophthalmol Vis Sci 1985;26:474–480.

11 Bazan HEP, Birkle DL, Beuerman R, et al: Inflammation-induced stimulation of the synthesis of prostaglandins and lipoxygenase-reaction products in rabbit cornea. Curr Eye Res 1985;4:175–179.

12 Bazan HEP: Corneal injury alters eicosanoid formation in the rabbit anterior segment in vivo. Invest Ophthalmol Vis Sci 1987;28:314–319.

13 Krause KH, Pittet D, Volpe P, et al: Calciosome, a sarcoplasmic reticulum-like organelle involved in intracellular Ca^{2+} handling by non-muscle cells: studies in human neutrophils and HL-60 cells. Cell Calcium 1989;10:351–361.

14 Stryer L: G-protein: A family of signal transducers. Annu Rev Cell Biol 1986;2: 391–419.

15 Axelrod J, Burch RM, Jelsema CL: Receptor-mediated activation of phospholipase

A_2 via GTP-binding proteins: Arachidonic acid and its metabolites as second messengers. Trends Neurosci 1988;11:117–123.
16 Braquet P, Touqui C, Shen TY, et al: Perspectives in platelet-activating factor research. Pharmacol Rev 1987;39:97–145.
17 Bazan HEP, Reddy STK: Production of platelet-activating factor (PAF) in the cornea after stimulation. Invest Ophthalmol Vis Sci 1989;30(suppl):404.
18 Bazan HEP, Reddy STK, Woodland JM, et al: The accumulation of platelet-activating factor in the injured cornea may be interrelated with the synthesis of lipoxygenase products. Biochem Biophys Res Commun 1987;149:915–920.
19 Bazan HEP, Braquet P, Reddy STK, et al: Inhibition of the alkali burn-induced lipoxygenation of arachidonic acid in the rabbit cornea in vivo by a platelet-activating factor antagonist. J Ocul Pharmacol 1987;3:357–365.
20 Bazan HEP, Bazan NG: Effect of BN-52021 on the arachidonic acid cascade in the inflamed cornea; in Braquet P (ed): Ginkgolides – Chemistry, Biology, Pharmacology and Clinical Perspectives. Barcelona, Prous Science, 1988 pp 769–773.
21 Raines EW, Dower SK, Ross R: Interleukin-1 mitogenic activity for fibroblasts and smooth muscle cells is due to PDGF-AA Science 1989;243:393–396.
22 Woodland J, Jumblatt MM, Neufeld A, et al: Epidermal growth factor (EGF) stimulates the hydrolysis of polyphosphoinositide and the accumulation of inositol triphosphate in cultured corneal endothelial cells. Invest Ophthalmol Vis Sci 1988; 29(suppl):112.
23 Squinto SP, Block AL, Braquet P, et al: Activation of HIV expression by PAF in human neuroblastoma cells. Trans Am Soc Neurochem 1989;20:135.
24 Squinto SP, Bazan NG: Induction of *c-fos* transcription by PAF in human neuroblastoma cells. Trans Am Soc Neurochem 1989;20:101.
25 Squinto SP, Block AL, Braquet P, et al: Platelet-activating factor stimulates a Fos/Jun/AP-1 transcriptional signaling system in human neuroblastoma cells. J Neurosci Res 1989;4:558.
26 Ashkenazi A, Ramachandran J, Capon DJ: Acetylcholine analogue stimulates DNA synthesis in brain-derived cells via specific muscarinic receptor subtypes. Nature 1989;338:146–149.
27 Maurice DM: The cornea and sclera; in Davson H (ed): The Eye, Orlando, Academic Press, 1984, pp 1–158.
28 Colley AM, Cavanagh HD, Law ML: Subcellular localization of muscarinic effects on enzymes of cyclic nucleotide metabolism in culture and corneal epithelial cells of the rabbit. Metab Pediatr Syst Ophthalmol 1987;10:36.
29 Limbird LE: Receptors linked to inhibition of adenylate cyclase: Additional signaling mechanisms. FASEB J 1988;2:2686–2695.
30 Sweatt JD, Schwartzberg MS, Frozer M, et al: Evidence for a role of Na^+/H^+ exchange in activation of human platelets by PAF. Circulation Res 1987;61:II(6)–II(11).
31 Sweatt JD, Conolly TM, Cragoe EJ, et al: Evidence that Na^+/H^+ exchange regulates receptor-mediated phospholipase A_2 activation in human platelets. J Biol Chem 1986;261:8667–8673.
32 Elgebaly SA, Downes RT, Bohr M, et al: Inflammatory mediators in alkali-burned corneas: Preliminary characterization. Curr Eye Res 1987;6:1263–1273.
33 Janoff A, Carp H: Proteases, antiproteases and oxidants: Pathways of tissue injury during inflammation. Monogr Pathol 1982;23:62–82.

34 Greenwald RA, Moz WW: Effect of oxygen-derived free radicals on hyaluronic acid. Arthritis Rheum 1980;23:455–463.
35 Fridovich I: Superoxide radical: an endogeneous toxicant. Annu Rev Pharmacol Toxicol 1983;23:239–257.
36 Rao GN, Broeckx JL, Church RL: Expression of c-myc proto-oncogene in primary cultures of human corneal stromal cells. Curr Eye Res 1989;8:405–414.
37 Allan G, Bazan NG: Differential increase in levels of *c-fos* and calcyclin, but not heat-shock protein, mRNAs in response to platelet-activating factor in a human neuroblastoma cell line. Soc Neurosci Abstr 1989;15(Part 1):842.
38 Rozsa A, Beuerman RW: Density and organization of free nerve endings of rabbit corneal epithelium. Pain 1982;14:105–120.
39 Harris LW, Purves D: Rapid remodeling of sensory endings in the corneas of living mouse. J Neurosci 1989:9:2210–2214.
40 Beuerman RW, Tanelian DL, Schimmelpfennig B: Nerve tissue interactions in the cornea; in Cavanagh HD (ed): The Cornea: Transactions of the World Congress on the Cornea III. New York, Raven Press, 1988, pp 59–62.
41 Beuerman RW, Schimmelpfennig B: Sensory denervation of the rabbit cornea affects epithelial properties. Exp Neurol 1980;69:196–201.
42 Mishima S: The effects of the denervation and the stimulation of the sympathetic and the trigeminal nerve on the mitotic rate of the corneal epithelium in the rabbit. Jap J Ophthalmol 1957;1:65–73.
43 Routtenberg A, Lovinger DM: Selective increase in phosphorylation of a 47KDa protein (F_1) directly related to long-term potentiation. Behav Neural Biol 1985;43: 3–11.
44 Ng SC, de la Monte SM, Conboy GL, et al: Cloning of human GAP-43: Growth association and ischemic resurgence. Neuron 1988;1:133–139.
45 Bazan HEP, Dobard P, Reddy STK: Calcium- and phospholipid-dependent protein kinase C and phosphatidylinositol kinase: Two major phosphorylation systems in the cornea. Curr Eye Res 1987;6:667–673.
46 Van Hoof COM, De Graan PNE, Oestreicher AB, et al: B-50 phosphorylation and polyphosphoinositide metabolism in nerve growth cone membranes. J Neurosci 1988;8:1789–1795.
47 Bazan HEP, Bazan NG: Composition of phospholipids and free fatty acids and incorporation of labeled arachidonic acid in rabbit cornea. Comparison of epithelium, stroma and endothelium. Curr Eye Res 1984;3:1313–1319.
48 Kikkawa U, Nishizuka Y: The role of protein kinase C in transmembrane signalling. Annu Rev Cell Biol 1986;2:149–178.
49 Scott BL, Bazan NG: Membrane docosahexaenoate is supplied to the developing brain and retina by the liver. Proc Natl Acad Sci USA 1989;86:2903–2907.
50 Thompson HW, Dupuy B, Beuerman RW, et al: Neural activity from the acutely infected HSV-1 rabbit cornea. Curr Eye Res 1988;7:147–155.
51 Bazan HEP, Sprecher H, Bazan NG: De novo biosynthesis of docosahexaenoyl phosphatidic acid in bovine retinal microsomes. Biochim Biophys Acta 1984;796: 11–19.

Nicolas G. Bazan, MD, PhD, LSU Eye Center and Neuroscience Center,
2020 Gravier Street, Suite B, New Orleans, LA 70112–2234 (USA)

Subject Index